MICROBIAL
PEST
CONTROL

BOOKS IN SOILS, PLANTS, AND THE ENVIRONMENT

Editorial Board

Soil Biochemistry, Volume 1, edited by A. D. McLaren and G. H. Peterson
Soil Biochemistry, Volume 2, edited by A. D. McLaren and J. Skujiņš
Soil Biochemistry, Volume 3, edited by E. A. Paul and A. D. McLaren
Soil Biochemistry, Volume 4, edited by E. A. Paul and A. D. McLaren
Soil Biochemistry, Volume 5, edited by E. A. Paul and J. N. Ladd
Soil Biochemistry, Volume 6, edited by Jean-Marc Bollag and G. Stotzky
Soil Biochemistry, Volume 7, edited by G. Stotzky and Jean-Marc Bollag
Soil Biochemistry, Volume 8, edited by Jean-Marc Bollag and G. Stotzky
Soil Biochemistry, Volume 9, edited by G. Stotzky and Jean-Marc Bollag
Soil Biochemistry, Volume 10, edited by Jean-Marc Bollag and G. Stotzky

Organic Chemicals in the Soil Environment, Volumes 1 and 2, edited by C. A. I. Goring and J. W. Hamaker
Humic Substances in the Environment, M. Schnitzer and S. U. Khan
Microbial Life in the Soil: An Introduction, T. Hattori
Principles of Soil Chemistry, Kim H. Tan
Soil Analysis: Instrumental Techniques and Related Procedures, edited by Keith A. Smith
Soil Reclamation Processes: Microbiological Analyses and Applications, edited by Robert L. Tate III and Donald A. Klein
Symbiotic Nitrogen Fixation Technology, edited by Gerald H. Elkan
Soil–Water Interactions: Mechanisms and Applications, Shingo Iwata and Toshio Tabuchi with Benno P. Warkentin

MICROBIAL
PEST
CONTROL

Sushil K. Khetan
Crop Protection Technologies Consultant
Pittsburgh, Pennsylvania

CRC Press
Taylor & Francis Group
Boca Raton London New York

CRC Press is an imprint of the
Taylor & Francis Group, an **informa** business

CRC Press
Taylor & Francis Group
6000 Broken Sound Parkway NW, Suite 300
Boca Raton, FL 33487-2742

First issued in paperback 2019

© 2001 by Taylor & Francis Group, LLC
CRC Press is an imprint of Taylor & Francis Group, an Informa business

No claim to original U.S. Government works

ISBN-13: 978-0-8247-0445-2 (hbk)
ISBN-13: 978-0-367-39801-9 (pbk)

Visit the Taylor & Francis Web site at
http://www.taylorandfrancis.com

and the CRC Press Web site at
http://www.crcpress.com

Foreword

Regulation of pesticides in the U.S. is a dynamic process being influenced from time to time by acts of Congress changing the basic laws. Significant changes were made in 1980 in the regulation and safety testing of naturally occurring pesticides ("biorational" pesticides). The United States Environmental Protection Agency (EPA) considers biochemical pest control agents, microbial pest control agents and microorganisms as biorational pesticides under the Federal Insecticide, Fungicide and Rodenticide Act (FIFRA), but its policy is to regulate only biochemical and microbial pest control agents. The Agency requires three types of data requirements–product analysis, hazard to humans and domestic animals (toxicology), and hazard to nontarget organisms–for the registration of biochemical and microbial pesticides. Several biopesticides, including microbial pesticides, have since been registered using these criteria.

Microbial pesticides have gained prominence in recent years for control of plant pests and diseases. The regulatory agencies have recognized these products, among others, as alternatives to conventional chemical pesticides which are known to show persistence in environmental media, lack specific mode of action, and leave toxic residues that affect human health and environmental safety. With the emergence of new technologies in agriculture, several new varieties of microbial products were developed and, especially during the last 5 years or so, the whole field of microbial pesticides has exploded.

Biotechnology is a broad term that encompasses a wide spectrum of applications used traditionally in the past to the rapidly advancing cutting edge technologies. It has had a profound impact on the food and agriculture industry. Some proponents of biotechnology see it as a way to greatly increase agricultural production while minimizing environmental damage. Some critics view biotechnology as causing devastation to the natural environment that will ultimately corrupt our ability to produce food. Lost in this exchange is the fact that biotechnology is a multifaceted spectrum of technological developments.

Biotechnology is currently the most powerful tool to further advance the various fields of agriculture. However, it can work practically only if it can be combined with established breeding strategies and with common agricultural practices. With the human population expected to reach 10 billion by the year 2050, the challenges facing agriculture and food research are enormous. Therefore, there is an urgent need to intensify our development of agricultural biotechnologies.

Transgenic crops are created by implanting specific gene sequences from a donor species with desired traits into a host plant to produce a new variety expressing the traits of the donor organism. For example, the gene sequences controlling the production of crystal endotoxins in *Bacillus thuringiensis* (*Bt*) bacteria have been transferred to cotton, corn, and potato plants to provide plant-wide insect resistance.

Many of the currently developed transgenic crops can, at least in the short term, enhance economic returns and decrease environmental impacts of agriculture. Whether this ultimately increases the sustainability of agriculture depends on whether the economic gains and environmental improvements continue in the future. Reduced reliance on chemical pesticides can benefit the surrounding ecology and the health of agricultural workers and consumers of agricultural products. However, if transgenic crops result in the development of resistant pest populations, the benefits of technology will disappear over time and may lead to increased chemical costs or reduced crop yields in the future. Similarly, transgenic crops that result in the creation of new weeds or diseases will not be sustainable.

The most widespread application of genetic engineering in agriculture by far is in engineered crops. Thousands of such products have been field tested and over a dozen have been approved for commercial use. The traits most commonly introduced into crops are herbicide tolerance, insect tolerance, and virus tolerance.

All of the commercially available insect-tolerant plants contain a version of the toxin *Bt*. A major concern is that wide use of *Bt* crops will lead to resistance to the toxin. If resistance develops, the *Bt* toxin will be useless as a pesticide. In such a case, the environmental benefits of the product will be short lived.

The EPA oversees genetically engineered microbial pesticides and certain genetically engineered crops under FIFRA. There appears to be two major concerns–pest resistance and safety of non-target insects–for transgenic crops such as transgenic cotton and corn, which have been approved for use in the U.S. These transgenic plants are developed by inserting genes from microbial pesticides that are known to be safe to humans and animals and do not harm beneficial insects. The EPA and USDA have been working with the industry to develop pest resistance management plans to avoid the build-up of resistance.

Another concern is the safety of the non-target organisms. In December 1999, EPA issued a notice to registrants of *Bt* corn plant-pesticides requiring submission of data on the effects of *Bt* corn pollen on non-target species, particularly the Monarch butterfly. This arose out of a preliminary study conducted by Cornell University suggesting that pollen from corn genetically-enhanced to express the *Bt* toxin may pose risks to Monarch butterfly larvae. To evaluate whether *Bt* corn pollen presents risks to the Monarch butterfly and other non-target lepidopteran species, the Agency requires extensive data including toxicity testing on the species of butterfly, milkweed distribution, and pollen dispersal and the development of appropriate models to test the risks to non-target insects in *Bt* corn fields.

Consumer acceptance is an important component of economic success of the biotechnology. If consumers decide that they prefer products without utilization of the genetically altered crops, then the producers of the genetically altered crops will face an uphill battle. Consumer perceptions of the risks and benefits associated with products derived from biotechnology appear to be driven by attitudes linked with specific applications of genetic engineering. It is essential to make the public understand the risks versus benefits of biotechnology.

In any country, there are people who want to apply new technologies including biotechnology in agriculture, and those who do not. The issues related to agricultural biotechnology seem to go deeper than religion or culture and they will always be divided. Bioethics addresses the ethical issues in biology and medicine using biotechnology. These ethical issues are not used to prescribe the correct answers but to make good choices for improving our life and society. The bioethical principles for agricultural biotechnology should consider, among many others, autonomy of choice versus justice, balancing benefits and risks, ethical values in life and nature, and sustainability and balancing ideals. Scientists should play a major role in bioethics of biotechnology and the public should be informed of the ethical, environmental and social issues that science raises.

The United States is rapidly moving in commercialization of genetically engineered products for crop production and crop protection. Introduced only a few years ago, the transgenic crops (cotton, corn and others) have already made significant inroads for agricultural biotechnology in the U.S. Potential markets are big; however, there are several unforeseen problems such as pest resistance, gene transfer to wild plants and unknown harms that may be realized over a long period. So long as the benefits from the new biotechnologies outweigh the potential risks, the U.S. is likely to promote the use of these new technologies for improved crop yields and reduced use of pesticides and fertilizers, among other economic, social and environmental benefits.

This book examines the state of the art, science and technology for the development, production, mode and mechanism of action, and application of bacterial and viral insecticides, biofungicides, bioherbicides and mycoinsecticides. Dr. Khetan has made an attempt to give an overview of the integrated uses of microbial pesticides with conventional chemical pesticides, and commercialization of microbial pesticides. Publication of this book seems timely when the regulators and the public are scrutinizing the transgenic crops and the genetically modified (GM) foods, while the industry is trying to convince them that GM foods are as safe as conventional foods and that transgenic crops would surely contribute to solving the global food problem in this century. This book will educate those who want to understand what these microbial pesticides are as well as the implications of their application to crop protection. Dr. Khetan should be commended for summing up the information in this field at this critical juncture.

N. Bhushan Mandava, Ph.D.
President
Mandava Associates

Preface

Growing concern for safety of human health and the environment has pushed the regulatory process to screen and control toxicants at levels that are perceived safe. There is growing worldwide regulatory effort to promote biologically based pest control that includes botanicals, biochemicals and microbials. Among these, microbial pest control agents as a group have seen rapid advances in recent years. The discovery and development of new microbial control agents and technologies capable of enhancing efficacy have been largely responsible for this. Genetic improvements of microbials using both recombinant and non-recombinant methods and compatibility with other interventions have resulted in a number of commercial products. The development and commercial availability of natural epizootics of fungi and viruses have expanded these options. The idea of using one microorganism against another, the predator versus the prey, offers an environmentally compatible and intellectually appealing approach for the control of plant diseases. Similarly, phytopathogenic fungi and bacteria have shown promise in weed management. A synopsis of the literature on the current uses of bacteria, viruses, and fungi as microbial control agents along with insights on their potential are presented in this book.

Microbials are not envisaged to replace chemical pesticides on major crops, such as cotton, corn, wheat and rice in the near future However, microbials are viewed as a key component of integrated pest management (IPM) programs, as supplements to chemical pesticides in combination or in succession. Selectivity and minimal environmental impact of microbials are desirable features for IPM programs. Also if they were used as direct replacements of chemical pesticides, then eventually these agents would face some of the same fate as the chemicals they replace, particularly with respect to resistance. Nevertheless, exclusive use of biopesticides may have niche markets on crops such as high-value fruits, nuts, vegetables and ornamentals. This is either as a consequence of some regulatory action or for tapping a specific activity unattainable through chemical pesticides.

There is a wealth of information on the state of the art and the level of commercialization reached among microbials for pest control. However, most of this is difficult to retrieve, being scattered in various discipline-based literature and Internet-accessible databases. Information retrieval of this nature also requires training in several specialized areas. The traditional chemical-based pest control products industry has long years of experience in developing and commercializing products for the plant protection market. However, the industry's technical pool, which consists of chemists and chemical engineers as well as entomologists and plant pathologists, faces difficulties in accessing and fully comprehending these new developments. On the other hand, the nascent small biologically-based pest control industry has struggled in devising product handling and commercialization strategies. These limitations may have constrained the growth and commercialization of biologically-based pest control agents. End-use based products integration has been in the offing within large chemical companies by forming life sciences divisions. However, medium and small chemical companies have been left out as they do not have the information resources of large companies, and have often been at a loss to figure out their areas of opportunity. There is an apparent need for a viable way for chemical and other specialty professionals for gathering first-hand knowledge of emerging microbial control systems and their potential.

My aim in writing this book was to bring together the latest advances in the science and technology of microbial pest control and in the evolving field of biopesticides. However this is not intended to be a comprehensive treatise. My effort is to present state of art information to the non-specialist reader in a manner the is easy and interesting. The approach is to facilitate understanding of this complex subject across discipline boundaries. A glossary of technical terms has also been prepared with this consideration. To aid rapid understanding of the material, numerous figures, drawings and chemical equations have been used. I hope these would also find use as teaching and presentation aids.

This book offers an overview of technologies that are driving the rapid proliferation of microbial based pest control. It also provides coverage of products that have reached the marketplace. Promising strategies for the development of effective biological controls for plant and vector borne pests, plant diseases and weed management are discussed by addressing many of the critical issues. Several current topics such as genetically altered Bacillus thuringiensis (*Bt*) and transgenic crops, microbial formulations and synergistic interactions of microbials with synthetic chemicals have been documented. Similarly, critical summaries on resistance management of *Bt* foliar applications and *Bt* genes in transgenic crops have been provided. Current states of technologies of viral and fungal insecticides and bacterial and fungal biofungicides and bioherbicides have been discussed. An analysis of related technical, social and economic issues that govern the commercialization efforts of biopesticides is also provided.

The book is divided into three parts consisting of eleven chapters. The first part covers various aspects of the *Bt* bacterium exclusively and consists of five chapters. *Bt* is the major success story of microbial pest control. The first chapter provides a comprehensive review of *Bt* delta-endotoxins. It also includes their mode of action as bioinsecticides and safety to non-target organisms and environment. The second chapter discusses *Bt* subspecies specific for control of caterpillar pests in agriculture and for control of mosquitoes that act as carriers of disease. A discussion on *Bacillus sphaericus* has been included, enumerating its growing commercial importance in vector control. Chapter Three covers various products based on genetically modified *Bt* as well as on transgenic *Bt* organisms and *Bt* plants. In Chapter Four, bacterial formulations including various efficacy-enhancing strategies are discussed. Various formulation types specific to *Bt* and *B. sphaericus* and their advantages are also covered. Chapter Five deals with resistance development in insect target populations towards *Bt* toxins by spray applications and through transgenic *Bt* plants. Resistance management strategies are also covered in this chapter.

Part Two consists of four chapters. Chapter Six discusses natural and recombinant viral bioinsecticides and the progress made in the use of baculoviruses for enhanced virulance and stability. Biofungicides are discussed in Chapter Seven, which covers fungal and bacterial antagonists, including their modes of action, and describes various commercial products. Fungal and bacterial pathogens used as bioherbicides are discussed in Chapter Eight. Chapter Nine on mycoinsecticides reviews the progress made in commercialization of entomopathogenic fungi. The third part consists of two chapters. Chapter Ten provides an overview of integrated use of biopesticides and use with synthetic pesticides. Several examples are provided here to demonstrate synergistic interactions of various combinations for reliable and consistent control of pests, weeds and diseases. The concluding Chapter Eleven discusses the issues involved in commercialization of biopesticidal products. It enumerates approaches employed for successful commercialization citing case studies from the literature.

It is hoped that industry managers, regulatory personnel and environmental safety and health planners and all those with an interest in safer approaches to pest control, will find the book a useful resource. This presentation will be found equally useful by crop and soil scientists, entomologists, plant pathologists, mycologists, microbiologists, formulation technologists, and biochemical engineers. As the field has grown rapidly, it may also provide useful overview of the state of art for researchers who are working on any single aspect of microbial pest control. It is my hope that the material contained here will lead to a better appreciation of microbial pest control strategies stimulating further research, production and commercialization of biopesticides.

Evolving sciences often progress by taking "two steps forward and one step backward." The ongoing controversy and intense speculation about AgBioTech industry are, perhaps, a manifestation of this axiom. It is observed that a market for a product exists based on the consumer's perception of it. Their level of awareness of the product shapes this perception. If this book enhances the level of awareness of microbial pest control by providing a small step towards understanding of many new developments, my objective in writing this book will have been achieved.

Acknowledgements

I acknowledge a large number of academic and industry scientists and technologists for their support. Particularly, I would like to thank Greg Boland (University of Guelph), Terry Couch (Becker Microbials), Brian Federici (University of California, Davies), Kunthala Jayraman (Anna University), N. Bhushan Mandava (Mandava Associates), Bineeta Sen (IARI, New Delhi) and Anurag Khetan (Merck & Co.) for offering their critical comments and useful suggestions during various stages of the writing. I acknowledge Ajay Gupta (Hindustan Insecticides), a valued colleague, who has provided active support throughout the preparation of the manuscript. His contributions towards literature search, preparation of review chapters on Bt, preparation of indices and proofreading of the manuscript, have been of immeasurable help. I also gratefully acknowledge the time and organizational support provided by the Chairman and Managing Director, Hindustan Insecticides Limited, New Delhi in completing a major part of this book.

Ashish Khetan (Carnegie Mellon University) provided significant technical support in the preparation of this manuscript and camera-ready copy. Kishan Lal did the initial typing. I would also like to thank the production team at Marcel Dekker, Inc., for providing consistent support and timely follow-up.

Finally, I would like to acknowledge the constant support of my wife Manju, who endured the endless hours of assimilating the information and putting it together in the manuscript. Without her patience and understanding this venture would not have been possible.

Sushil K. Khetan, Ph.D.

Contents

Part I

Bacterial Insecticides

1

Bacterial Insecticide:
Bacillus thuringiensis

1 Introduction

The critical need for safe and effective alternatives to chemical insecticides has stimulated considerable interest in using pathogens as biological control agents for insects of agricultural and medical importance. Both chemical and biological insecticides are currently used for insect control but the use of biological insecticides is favorable because they kill undesirable agricultural and household pests and vectors of human and animal diseases without introducing toxic and non-biodegradable substances into the ecosystem.

Microbial pesticides are becoming recognized as an important factor in crop and forest protection and in insect vector control. These pesticides are naturals, disease causing microorganisms such as viruses, bacteria and fungi, that infect or intoxicate specific pest groups. The pathogen that has been the most successful and holds considerable potential for further development is the insecticidal bacterium *Bacillus thuringiensis (Bt)*. Other bacterial insecticides exist, but have not yet achieved the commercial success of *Bt*.

Bt-based insecticides provide growers and the public with environmentally friendly and effective alternatives to conventional insecticides. These are (a) not harmful to natural enemies and other non-target organisms due to narrow host specificity, (b) harmless to vertebrates including mammals,

3

(c) biodegradable in the environment and (d) highly amenable to increase in activity by genetic engineering.

A number of excellent reviews have appeared in the literature on the subject (Schnepf et al., 1998, Prieto-Samsonov et al., 1997; Cannon, 1996; Kumar et al., 1996; Koziel et al., 1993; Beegle and Yamamoto 1992; Lambert and Peferoen 1992 and Peferoen, 1992).

1.1   History

The first record on *Bt* goes back to 1901, when Ishiwata discovered a bacterium from diseased silkworm larvae that he named *Bacillus sotto*. Berliner isolated a similar bacterium from Mediterranean flour moth (*Anagasta kuehniella*) in 1911. The bacterium was named *Bacillus thuringiensis* after Thuringia, Germany, where the infected insect was found.

Husz isolated a *Bt* strain from *Ephestia* in 1928, tested it on European corn borer and reported the first application of *Bt*. This work eventually led to the first commercial product, Sporeine, which was produced in France in 1938. The development of potent organic insecticides, however prevented rapid advancement in the development of biological alternatives for pest control. The pioneering research of Steinhaus (1951) on *Bt* and a growing realization that organic insecticides are deleterious to the environment and human health spurred renewed interest in *Bt*. This led to the introduction of a viable *Bt* biopesticide called Thuricide in 1957, a product name still in use.

During 1960s, several commercial formulations of *Bt* were manufactured in the United States, Soviet Union, France, and Germany and used with various degrees of commercial success. A major advance occurred when Howard Dulmage and Clayton Beegle of the USDA Agricultural Research Service assembled the first collection of *Bt* strains. Dulmage isolated the HD-1 strain, which is still used as a reference in many studies and is included in many *Bt*-based products (Dulmage, 1970).

Until the 1970s, it was generally believed that *Bt* was only active against Lepidoptera. The discovery of *Bt* subsp. *israelensis*, which is toxic to larval mosquitoes and blackflies (Goldberg and Margalit, 1977) and the discovery of *Bt* subsp. *tenebrionis* (Krieg et al., 1983), which is toxic to several beetle species, stimulated sharply increased commercial interest in *Bt*. During the 1980s, it led to foundation of new biotech companies whose core business consisted of the development of novel *Bt*-based insecticides. By 1992, it was estimated that 40,000 strains of *Bt* were stored, mainly in private collection worldwide. Now, numerous commercial products formulated with *Bt* are registered for control of larvae of lepidopterous insects (moths and butterflies), dipterous insects (flies and mosquitoes) and coleopterous insects (beetles).

1.2   Occurrence

Bt is ubiquitous in soils and the foliar surface (phylloplane). These habitats must in addition to insect larvae, constitute viable niches of *Bt*. Indeed, *Bt* is a facultative pathogen, in contrast to *B. popilliae* for example, which is strictly parasitic and cannot live independent of its host.

   Bt has a worldwide distribution in various soils, including forest soils, savannas and deserts and in insect rich environments such as sericulture farms, insect-rearing facilities, grain dust from flour mills, and grain storage facilities. The environmental distribution of *Bt* is not related to the distribution of the target insect(s). Nevertheless, certain *Bt* serotype appear to accumulate in the environments where insects are abundant and/or breeding (Delucca et al., 1981; Bernhard et al., 1997 and Chaufaux et al., 1997).

   Bt can be isolated from a wide variety of habitats. It is readily found in cadavers of insects from stored product dusts than from soil, but it doesn't grow well in many of these because it is not a major or dominant species (Bernhard et al., 1997; Chaufaux et al., 1997). Moreover, it is interesting that, unlike many viruses, fungi and protozoa, *Bt* has never been reported as the cause of large-scale epizootics. Despite *Bt*'s wide spread occurrence in nature, optimal reproduction probably occur in insects (Federici, 1993).

   Bt is easily grown and completes its growth cycle on ordinary laboratory media with a minimum of nutrients. It is, however, incorrect to assume that *Bt* germinates, grows and multiplies in nature if appropriate nutrients are available. Although *Bt* spores may survive for many years, they do not germinate and multiply as vegetative cells in natural soils. This observation indicates that *Bt* is not adapted to the soil environment. Under natural conditions, crystal proteins released from the spores into soil are rapidly degraded, but spores are more long lived. Although these spores have some toxic activity, extremely high numbers of spores are required to kill larvae. Therefore, given the low number of *Bt* spores, this bacterium has probably no significance in the control of the insect populations in natural environments and is unable to multiply in this environment (Saleh et al., 1970 and Lambert and Peferoen, 1992).

1.3   Characteristics

Bacillus thuringiensis (*Bt*) is a gram-positive, rod shaped, aerobic, endospore forming, ubiquitous soil bacterium that belongs to morphological group I, along with *Bacillus cereus, Bacillus anthralis* and *Bacillus laterosporus*.

   Bt produces a parasporal body within its cells during sporulation. This parasporal body contains one or more proteins, typically in a crystalline form

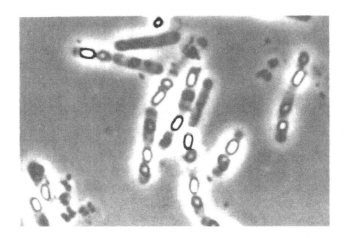

Figure 1.1   Photomicrograph of *Bacillus thuringiensis* bacteria. The white spots inside the bacteria are the protein crystals (Courtesy Monsanto Co.).

(Figure 1.1). These proteins possess insecticidal properties and are called *Bt* toxins, δ-endotoxins or insecticidal crystal proteins (ICPs). Insecticidal crystal proteins produced by *Bt*, kill insects by binding to and disrupting their midgut epithelium membranes, yet does little or no harm to most other organisms, including human, wildlife and other insects.

Bt produces a large variety of insecticidal crystal proteins, which reflects its remarkable adaptive capacity to settle in the large ecological niche constituted by insects. The biological mechanisms used by the bacterium to make enough toxin to kill insects are a striking example of bacterial evolution. The insecticidal proteins are synthesized during the stationary phase and accumulate in the mother cell as a crystal inclusion which account for 20 to 30% of the dry weight of the sporulated cell and mainly consists of protein (95%) and a small amount of carbohydrate (5%). The amount of the crystal protein produced by a *Bt* culture in laboratory conditions (about 0.5 mg of protein/ml) and the size of the crystal indicate that each cell synthesizes 10^6 to 2×10^6 δ-endotoxin molecules during the stationary phase to form a crystal. This would require a substantial dedicated usage of cell machinery. Nevertheless, sporulation and the associated physiological changes also proceed in parallel with δ-endotoxin production (Agaisse and Lereclus, 1995). The δ-endotoxins, which are deposited in the cytoplasm of the bacterium, are of increasing importance as biopesticides. In this respect, *Bt* represents an intermediate between biocontrol and synthetic insecticides, with a role restricted to that of an ingestion toxicant (Cannon, 1996).

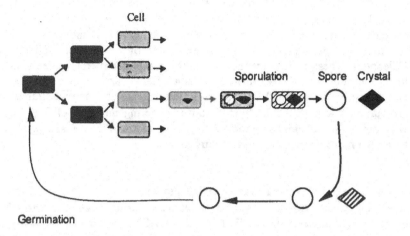

Cell

Sporulation   Spore   Crystal

Germination

Figure 1.2 Schematic representation of a sporulating *Bacillus thuringiensis* (Adapted from Lambert and Peferoen, 1992).

1.3.1 Life cycle

Bt has two growth phases, germination (vegetative growth) and sporulation (Figure 1.2). During the vegetative cell cycle, when the bacteria are in nutrient rich environment, *Bt* normally multiplies exponentially by cell division. Sporulation occurs when nutrients are depleted or when the environment become adverse, forming spores (called endospores) within a sporangium. These spores are highly resistant to adverse conditions such as heat and drought, enabling the bacterium to survive periods of stress. Spores germinate and may restart a vegetative cell cycle under favorable conditions (Lambert and Peferoen, 1992 and Itoua-Apoyolo et al., 1995).

1.3.2 *Toxins produced by* Bt

Bt produces mainly δ-endotoxins and a wide variety of secondary metabolites, some of which are insecticidal e.g. the heat stable adenine nucleotide, β-exotoxin and immune inhibitor A (protease), several phospholipases and a heat-labile α–exotoxin, lecithinase C. More recent characterization has shown that chitinases (Sonngay and Panbangred, 1997), zwittermicin (Stabb et al., 1994) which increases the toxicity of Cry proteins by unknown mechanism (Federici, 1998), and the secreted vegetative insecticidal proteins may contribute to virulence (Estruch et al., 1996).

a) δ-Endotoxin   The insecticidal crystal preoteins (ICPs) or δ-endotoxins are a family of insecticidal proteins produced by *Bt* during sporulation. The δ-endotoxins have relative molecular masses of 60–70 kDa in the active form and specific toxicities against insects in the order of Lepidoptera, Diptera and Coleoptera. These toxins have been formulated into commercial insecticides for three decades and now insect-resistant plants are engineered by the transformation with Lepidoptera-specific toxin genes. Numerous strains of *Bt* are currently known. Each strain produces differing numbers of δ-endotoxins with various insecticidal activities. A given δ-endotoxin is typically insecticidal towards a narrow spectrum of insect targets.

b) β-Exotoxin: Several strains of *Bt* produce low molecular weight heat stable insecticidal toxin known as β-exotoxin during vegetative growth. β-exotoxin or thuringiensen ($C_{22}H_{32}N_5O_{19}P_1$. 3 H_2O) is water soluble, dialyzable nucleotide composed of adenosine, glucose and allaric acid (Figure 1.3). It is found that the phosphate group on allaric acid is essential for activity and that the toxicity of β-exotoxin results from its inhibition of RNA biosysthesis by competing with ATP for binding sites. Some of the *Bt* subsp. that produce β-exotoxin are *canadensis, darmastadiensis, kumamotoensis, galleriae, morrisoni, thuringiensis* etc.

Two β-exotoxin containing products are produced and used in Russia, Thuringin 1 and 2 (2% and 10% β-exotoxin respectively) and Biotoxibacillin (0.6 - 0.8% β-exotoxin). The products are used effectively against several species of red mites, as well as against larvae of houseflies and blowflies. Other products are Dibeta (Abbott labs.) and Exobac (CRC). Current registration standards for *Bt* based insecticides requires an absence of β-exotoxin (Cambell et al., 1987 and Prieto-Samsonov et al., 1997).

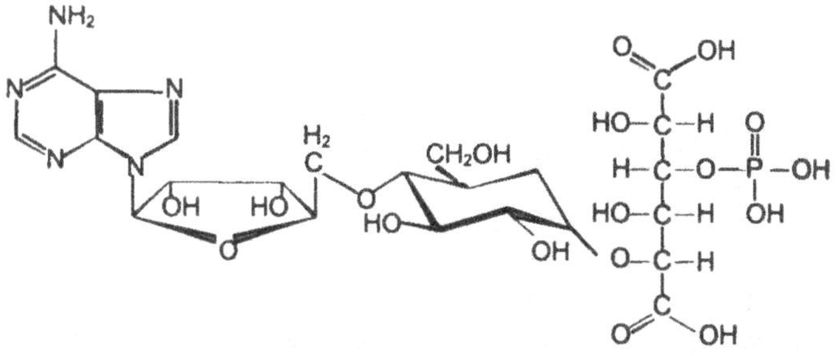

Figure 1.3   β-Exotoxin.

c) α-Exotoxin: Many strains of *B. thuringiensis* produce heat-labile insecticidal exotoxin known as α–exotoxin and found to be toxic to wax moth larvae. α-exotoxin is also toxic to insect and mice upon injection. α-Exotoxin extracted in water from the Thuricide powder, a commercial preparation of *Bt*, was toxic to sawfly larvae fed on treated foliage. The estimated size of α-exotoxin was 45-50 kDa (Beegle and Yamamoto, 1992). Some of the *Bt* subspecies that produce α-exotoxin are *aizawai, alesti, finitinus, kenyae, kurstaki, sotto, thompsoni, thuringiensis* etc.

d) Vegetative Insecticidal Proteins (Vips): The clarified culture supernatant fluids collected during vegetative (log phase) growth of *Bacillus* species are a rich source of insecticidal activities. Vip proteins are distinct from insecticidal proteins that belong to δ-endotoxin family, which are ineffective against rootworm and cutworms (Nunez-Valdez, 1997). These Vips do not form parasporal crystal proteins and are apparently secreted from the cell.

Vips display acute biological activities against major pests of maize, such as the corn rootworm (*Diabrotica* spp.) and the black cutworm (*Agrotis ipsilon*). Supernatant fluids of certain *B. cereus* isolates possess acute bioactivity against Northen and Western corn root worms. The insecticidal properties are associated with a binary system, whose two proteins are termed Vip1A and Vip2A. Similarly, supernatant fluids of certain *Bt* cultures have potent insecticidal properties against black cutworm and fall armyworm. This led to discovery of Vip3A, a novel insecticidal protein with no homology with any known protein. This protein is reported to exhibit toxicity towards a wide variety of lepidopteran insects, including *Agrotis ipsilon, Spodoptera exigua, Spodoptera frugiperda* and *Helicoverpa zea* (Estruch et al., 1996).

2   Production of *Bt*

Bacillus thuringiensis has been used to produce biological insecticides for foliar application for over 50 years. Bioinsecticides derived from *B. thuringiensis* contain inclusion bodies, spores, cell debris and other residual solids, which are all recovered from the broth at the end of the fermentation and then formulated, packed and dried (Rowe and Margaritis, 1987).

Bioinsecticides were, and are still produced by fermentation of single *Bt* strains in crude, inexpensive media. As the *Bt* cells begin to exhaust the medium, they enter the sporulation phase of growth where one or more insecticidal proteins, or δ-endotoxins, are synthesized. The δ-endotoxins accumulate in morphologically distinct crystal(s) within the cells and are released into the medium with the spores when the cell ultimately lyse.

Typically, the first step in the manufacturing process of a *Bt* bioinsecticide is to concentrate the fermentation solids (spores, δ-endotoxin crystals and unused particulate media components) by centrifugation. Centrifugation or filtration can recover the suspended solids from the fermented broth of *Bt* with efficiency higher than 99% based on spore count (Villafana-Rojas et al., 1996). Sometimes tangential flow dialysis is also used to further concentrate the solids before spray drying. The dried material is then milled to a uniform size to produce a technical powder that can be formulated in a variety of ways. The most common formulations are wettable powders or dispersible granules, but oil and aqueous flowable preparations have also been produced.

Three major factors contribute to the effectiveness of *Bt* insecticides. First, a *Bt* strain must be obtained that produces δ-endotoxins which are effective on target insects. Second, the strain must be able to produce large amounts of active δ-endotoxin during large-scale fermentations at a minimum cost. Third, a formulation that maximizes the longevity and effectiveness of the δ-endotoxins should be used (Kozeil et al., 1993).

2.1   Strain Selection

The bacterial strains used for all of the early and some of the current *Bt* foliar insecticides are wild type strains, that is, they were found in nature in the form in which they are used to produce the microbial spray. Most *Bt* isolates produce multiple δ–endotoxins, each having a characteristic insecticidal activity spectrum. It is generally believed that the overall activity spectrum of a strain is due to additive and/or synergistic interspectrum of the individual δ-endotoxins present in their proportional amounts. The maximum amount of a single δ-endotoxin protein would be obtained from strain expressing a single δ-endotoxin gene at its maximum biosynthetic capacity. However, the insecticidal spectrum of such a strain would be relatively narrow. In contrast, a strain producing several δ-endotoxins would have a broader spectrum, but the activity contribution of each δ-endotoxin would be decreased in proportion to the amount produced.

It is possible to construct strains with increased activity on specific pests by using genetic manipulation techniques and genetic engineering to eliminate unnecessary δ-endotoxin and/or introduce desirable ones. *Bt* strains present in several currently available bioinsecticides were constructed using these techniques e.g. Agree (Thermo Trilogy), Novodor (Abbott), Condor, Foil and Cutlass (Ecogen) etc. (See Chapter 3).

The task of identifying δ-endotoxins with new or potent activities typically involves screening wild type strain of *Bt*. When screening for specific activity, is some times the case that toxicity to a target species is greatest in *Bt* isolates which commonly occur in the same microbial habitat as the pest. For example, the majority of *Bt* isolates collected from silkworm litters were toxic to silkworm larvae, whereas most isolates from soil collected in Japan showed no insecticidal activity against this species (Koziel et al., 1993).

A large number of *Bt* isolates are screened in an attempt to identify new strains with increased levels of insecticidal activity and activities against a broader spectrum of insect pests. The discovery rate of novel *Bt* strains or ICPs is relatively high, up to 1 in 1000 screened *Bt*s produce a novel ICP compared to 1 in 20,000 screened compounds for synthetic pesticides. The activity spectrum of *Bt* is still expanding. Thus, *Bt* may provide an alternative to many currently used chemical pesticides (Lambert and Peferoen, 1992). Various methods are used to identify the novel activities or rank the relative potencies of *Bt* strains. These include bioassay, molecular probe using gene specific polymerase chain reactions (PCRs) and other.biochemical or immunological methods.

2.2 Fermentation

The development of an economical fermentation process involves a great deal of experimentation and is largely empirical and must be specifically designed for a particular strain of *Bt*. Thus the production of a new bio-insecticide using a new *Bt* strain requires the development of a different fermentation process. This may only require minor changes in an existing process or the physiological properties of the new strain may be different enough to require major changes in the entire manufacturing process (Dulmage and Rhodes, 1971 and Koziel et al., 1993). Although strains of *Bt* differ, they all share some common properties. It is therefore possible to define some basic requirements for production of *Bt*. As production of δ-endotoxin occurs only during sporulation, optimal sporulation is essential. Under anaerobic conditions, growth of *Bt* is slow and sporulation may be inhibited; good oxygen supply is therefore an essential requirement. High frequency plasmid loss was induced in *Bti* by cultivating it at 43°C (Gonzalez and Carlton, 1984). Therefore, cultivation at higher temperature is best avoided. In case of recombinant *Bt* containing plasmids, typically, temperature during production is kept at 28°C. *Bt* is not particularly sensitive to pH and optimum growth occurs between pH 6.5 and pH 7.5 (Bernhard and Utz, 1993).

The cost of production is one of the major issues in commercializing biopesticides products. The most important factors that determine the cost of production are raw material cost and the product yield. In the recent past, various raw materials of agriculture origin and industrial residues have been investigated to replace the expensive ingredients currently used in the manufacture of *Bt*. The cost of these materials vary between 20 to 60% of the overall cost of production. These raw materials are wheat bran, rice bran, defatted soybean cake, powder of leguminus seeds, molasses, corn steep liquor, fermented cassava, ground and whole maize, fish meal etc. The raw material cost can be minimized by selecting substances which are locally available on year round basis, have minimum environmental risks, sustain high specific growth rate, sporulation and toxin yield and have minimum cost (Tyagi, 1997).

Bt can be produced both in semi-solid and in submerged fermentation, both of which are used commercially. However, the details of the fermentation, downstream processing and formulation of *Bt* products are generally kept as trade secrets.

a)   Semi-solid fermentation: Mechalas patented a process in 1963 for the large-scale production of *Bt* toxins by surface cultures on bran-based semi-solid media. In a typical production with semi-solid media, nutrients are contained in a coarse porous matrix with a large surface area through clumping induced by excessive moisture. In large cultures, oxygen transfer, humidity and temperature are maintained by either placing the media on shallow trays, or keeping them in aerated drums. Dulmage (1983) has reviewed semi-solid production of *Bt*.

b)   Submerged fermentation: A typical *Bt* fermentation (Figure 1.4) is started by inoculating a flask of culture medium with a loopful of bacteria from an agar plate. The flask is then incubated at 28°C on a shaker. At this stage, spores germinate and the vegetative cells adapt to the culture medium. After 10-16 hours, the cells are in mid-log phase. The culture is then used to inoculate a seed fermenter containing the same culture medium. Counting cells microscopically monitors growth in the fermenter. After 6-8 hours, the culture is in the late log-phase and is transferred into the larger fermenter. If the transfer is not delayed too long, the cells will continue to grow exponentially. Growth in the fermenter is again monitored microscopically until the sporangia start to lyse.

There are advantages and disadvantages in each type of fermentation. Production on semi-solid media allow cultivation of microorganisms which either do not grow at all or fail to produce the desired product in submerged culture. Semi-solid media are, however difficult to sterilize and to keep sterile during production. Maintaining defined, uniform conditions during the production process as well adjusting parameters like pH, is also difficult. Therefore, the deep liquid fermentation, in which growth of bacterium takes place in liquid medium in large mechanically agitated vessels sparged with air, is the preferred method of production for current *Bt* insecticides (Bernhard and Utz, 1993).

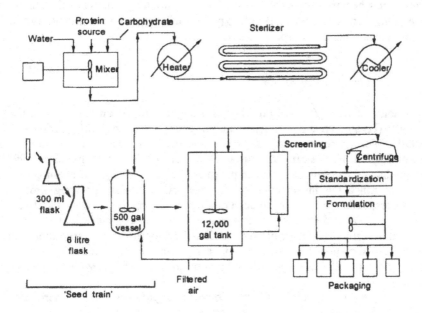

Figure 1.4   Submerged fermentation of *B. thuringiensis*

2.3   *Formulation*

Bt bioinsecticides have been formulated in many ways since their introduction in 1938. Wettable powders, dispersible granules, dust, microgranules, aqueous flowable liquids and oil based flowable liquids that are dispersible in water have been developed (See chapter 4). Early formulations were often difficult to apply and their performance was often poor or unreliable. Advances in application equipment and improvement in formulation technology have largely overcome application difficulties. However, different strains may well require different formulation because of difference in their respective fermentation media and their by products. Also, batch to batch formulation problems can be encountered due to variable quality of the fermentation components.

3   Classification and Nomenclature of *Bt*

As *Bt* became increasingly commercially important and numerous different isolates were discovered, the need for a method to identify and classify *Bt* subspecies becomes apparent. They can be classified by a number of techniques including serotyping, crystal serology, crystal morphology, protein profiles, peptide mapping, DNA probing and insecticidal activity.

3.1   Serotyping of Bt Strain

Vegetative cells of *Bt* have (at least) two major antigens on their surfaces: the flagella (H) antigen and the heat-stable somatic (O) antigen (HSSA) (Mike et al., 1990). The H-serotyping has been succesful in the classification and identification of *Bt* strain (de Barjac, 1981 and de Barjac & Francon, 1990). More than 40 *Bt* serotypes are recognised on the basis of flagellar H-antigens.

Since the introduction of this method, a rapidly growing group of serotypes, called serovars have been described. The method is now widely used by researchers and officials to designate and distinguish individual *Bt* strains. Although many researchers use the word 'variety' for these taxa, according to the international code of Nomenclature of Bacteria, the correct name is 'subspecies',. Examples are *B. thuringiensis* subsp. *tenebrionis*, *B. thuringiensis* subsp. *aizawai* etc.

Despite the fact that the biological significance of the serotype is unclear, the method is still used for general classification of *Bt*. However, the serotype has certainly very little to do with the insecticidal activity or presence of certain crystal proteins. For example, *Bt* subsp. *tenebrionis* shares the same antigens as serovar (=variety) *morrisoni* (8a8b) and can only be distinguished in terms of their pathogenicity to coleopteran larvae (de Barjac and Francon, 1990). There is no immunological relationship between the crystal toxin of *Bt* subsp. *tenebrionis* and δ-endotoxin of the other strains in H-serotype 8a8b. Further more, *Bt* subsp. *tenebrionis* belongs to O-serotype IX (in contrast to strain PG-14) (Krieg et al., 1987). Thus, subsp. *tenebrionis*, which is very unique in host range crystal morphology, crystal toxic gene and crystal protein chemistry, does not merit separate status. Moreover, several strains producing the same crystal proteins with same spectrum of activity belong to different serotype. Similarly, not all serover *aizawai* strains are active against the cotton leaf worm (*Spodoptera littoralis*). The *morrisoni* serotype harbors strains that are active against Lepidoptera (strain HD-12), Diptera (strain PG-14) and Coleoptera (*Bt* subsp. *tenebrionis*). The recent techniques that have been applied to the identification/classification of *Bt* are high performance liquid chromatography (HPLC), plasmid mapping and cloning and sequencing of crystal toxin genes.

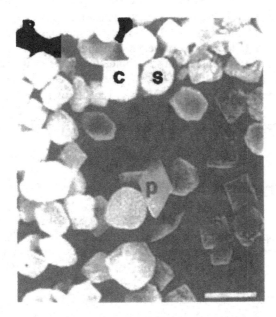

Figure 1.5 *Bt* crystal inclusions include bipyramidal, spherical and rectangular shapes (Source: Lopez-Meza et al., 1995)

3.2 *Morphology of Parasporal Inclusions*

In general, the shape of the parasporal body is a good but not absolute indication of an isolate's pathotype. *Bt* isolate produce water-insoluble, parasporal inclusions of a wide range of morphological types. Crystal inclusion can be classified into various types, such as bipyramidal (Cry1), cuboidal (Cry2), flat rectangular (Cry 3A), irregular (Cry3B), spherical (Cry4A and Cry4B) and rhomboidal (Cry11A) (Figure 1.5).

Bipyramidal crystals show a greater frequency of toxicity than all other types, and the majority of isolates with lepidopteran activity contain such inclusions. Irregular spherical crystals can be mosquitocidal, rhomboid are often active against certain coleopteran species and irregular pointed (eg.triangular) crystals include those with no identified toxicological activity. Bipyramidal crystals synthesized in host cells are typically about 1.1μm long and 0.5μm wide. *Bt* subsp. *israelensis* (H-14) and *Bt* subsp. *morrisoni* (PG-14 isolate), produce proteins that crystallize in different shapes such as spherical parasporal body (0.7-1μm) that is toxic primarily to nematocerous dipterans, e.g. mosquito and blackfly (Federici et al., 1990). *Bt* subsp. *tenebrionis* produces a parasporal crystal that is toxic to certain species of Coleoptera. The crystal appears as flat plates with square, rectangular or rhomboidal outlines, their axes measures 0.7-2.4 x 0.7μm and the thickness of the plates ranges from 0.15-0.25μm with pointed or roof shaped edges (Keller & Langerbruch, 1983). The degree of complexity within the parasporal bodies can be composed of single type of protein molecule or a mixture of as many as three (Lopez-Meza and Ibarra, 1996).

4   *Bt* Crystal Proteins and Genes

Crystal proteins have been classified on the basis of their insecticidal and molecular properties and are assigned capitalized three letter Roman Code, Cry. The corresponding gene is assigned three letter lower case italicized code *cry*.

4.1   *Classification and Nomenclature*

Based on the primary target insect specificity and sequence similarity, the insecticidal crystal proteins were classified by Hofte and Whiteley (1989) into five major classes; four Cry classes, Cry (-I, -II, -III and –IV) and a Cyt class (Cyt for cytolytic crystal protein). The roman numeral indicated pathotype (I for toxicity to lepidopterans, II for both lepidopterans and dipterans, III for coleopterans and IV for dipterans in the suborder Nematocera). Feitelson et al. (1992) added two new major classes of nematode-active toxins, CryV and CryVI to the Hofte and Whiteley classification. The CryV name was given to protein toxic to both lepidopterans and coleopterans. The crystal protein belonging to each of these classes were grouped in subclasses (A.B.C....and a, b, c.....) according to sequence. The upper case letter indicating the chronological order in which genes with significant differences were described, the lower case letter in parentheses, indicated minor differences in the nucleotide sequence within a gene type. Thus, CryIA refers to protein toxic to lepidopterous insects for which the first gene (*cryIA*) was sequenced, whereas CryIVD refers to the protein for which the encoding gene was fourth from this pathotype.

Crickmore et al. (1995) introduced a nomenclature for classifying the *cry* genes and their protein products on the basis of comparative amino acid sequence identity of the full-length gene products. Many *cry* genes retain the name assigned by Hofte and Whiteley with a substiution of Arabic for Roman numerials (eg. *cryIA(a)* become *cryIAa* (Table 1.1 and 1.2).

Table 1.1 Homology of amino acids of *cry* genes

Rank	Symbol	% Homology of amino acid of cry gene
Primary	Arabic no.	45%
Secondary	Upper case letter	75%
Tertiary	Lower case letter without parentheses	95%
Quaternary	Allele number	95%

Table 1.2 Nomenclature of *Bt* insecticidal protoxins (Adapted from Crickmore et al., (1999)

Insecicidal protoxins		Molecular Mass (kDa)	Toxicity
Hofte and Whiteley (1989)	Crickmore et al. (1999)		
CryIA(a)	Cry1Aa	133.2	Lepidoptera
CryIA(b)	Cry1Ab	131.0	Lepidoptera/Diptera
CryIA(c)	Cry1Ac	133.3	Lepidoptera
CryIA(d)	Cry1Ad		
CryIA(e)	Cry1Ae		
CryIB	Cry1Ba	138.0	Lepidoptera
CryIC(a)	Cry1Ca	134.8	Lepidoptera/Diptera
CryIC(b)	Cry1Cb	134.0	Lepidoptera
CryID	Cry1Da	132.5	Lepidoptera
CryIE	Cry1Ea	132.0	Lepidoptera
CryIE(b)	Cry1Eb		
CryIF	Cry1Fa	133.6	Lepidoptera
CryIG	Cry9Aa	130.0	Lepidoptera
CryIX	Cry9Ba	81.0	Lepidoptera
CryIH	Cry9Ca	129.9	Lepidoptera
CryIIA	Cry2Aa	70.0	Lepidoptera/Diptera
CryIIB	Cry2Ab	70.8	Lepidoptera
CryIIC	Cry2Ac	71.0	Lepidoptera

Table 1.2 (continued)

Insecicidal protoxins		Molecular Mass (kDa)	Toxicity
Hofte and Whiteley (1989)	Crickmore et al. (1999)		
CryIIIA	Cry3Aa	73.1	Coleoptera
CryIIIB	Cry3Ba	74.2	Coleoptera
CryIIIB2	Cry3Bb		
CryIIIC	Cry7Aa	73.0	Coleoptera
CryIIIC(b)	Cry7Ab	73.0	Coleoptera
CryIIID	Cry3Ca	73.0	Coleoptera
CryIIIE	Cry8Aa		
CryIIIF	Cry8Ca		
CryIIIG	Cry8Ba		
CryIVA	Cry4Aa	134.4	Diptera
CryIVB	Cry4Ba	127.8	Diptera
CryIVC	Cry10Aa	77.8	Diptera
CryIVD	Cry11Aa	72.4	Diptera
CryV	CryIIa	80.0	Lepidoptera/Coleoptera
CryVB	Cry12Aa		
CryVC	Cry13Aa		
CryVD	Cry14Aa		
CryVIA	Cry6Aa	54.0	Nematode
CryVIB	Cry6Ba		
CytA	Cyt1Aa	27.4	Diptera/Cytolytic
CytB	Cyt2Aa	29.0	Diptera/Cytolytic

4.2   Bt cry Genes:Encoding Protoxins

Bt toxin genes are usually plasmid born but also chromosomally, they are encoded on the transmissible plasmid of molecular weight of 40-150 MDa (Kumar et al., 1996). The protoxins encoded by *cry* genes are processed into the toxin fragments in the midgut of insect and these in turn are responsible for insecticidal activity.

The *cry* genes encode insecticidal proteins, either of 130 to 140 kDa or of ca. 70 kDa. For the 130 to 140 kDa proteins, the toxic segment is localized in the N-terminal half of the protoxin. The C-terminal part of the ca. 130 kDa proteins (Cry1, Cry4a, and Cry4B) is not essential for toxicity, but is the most highly conserved domain of these crystal proteins. The Cry2A, Cry3 and Cry11A (formerly designated as CryIVD) protoxins are smaller.

4.3 Cry Gene Types

a) *cry1* gene: Genes of the class *cry1* (eg. *cry1A, -B, -C* etc.) encode 130-140 kilodaltons (kDa) in molecular mass proteins which accumulate in bipyramidal crystalline inclusions formerly termed P1 crystals during the sporulation of *Bt*.

The Lepidoptera specific crystal proteins are the best studied crystal proteins (Masson et al., 1990). Cry1A is the most common type of crystal protein that is found in Cry1 producing *Bt* strain and *cry1Ab* gene appears to be most widely distributed gene amongst *Bt* subspecies. Cry1Aa is normally highly toxic for larvae of the silkworm, *Bombyx mori*, whereas Cry1Ac is virtually nontoxic. Cry1C δ-endotoxin is toxic to the larvae of the *Spodoptera* species of the order Lepidoptera and is one of very few δ-endotoxin active against these polyphagous insects (Smith et al., 1996).

b) *cry2A* genes: The *cry2A* gene encode 65kDa proteins (formely known as P2 crystal protein), which occur as distinct cuboidal inclusions in strains of several *Bt* subspecies *kurstaki* (HD-1, HD-263, NRD-12 and 14 other strains), *thuringiensis, tolworthi* and *kenyae*. The Cry2Aa protein is toxic to both lepidopteran and dipteran larvae, whereas Cry2Ab toxin is only toxic to lepidopteran insects.

c) *cry3* genes: *cry3* genes encode 72 kDa protein which form flat rhomboidal crystals in *Bt* subsp. *tenebrionis, sandiego* etc. The Cry3A protein displays coleopteran toxicity. For most ICP genes transcription is concomitant with sporulation, however Cry3A protein appears to be unique in being expressed during vegetative growth of *Bt* cell.

d) *Bti cry* genes: The *cry4* class of crystal protein genes is composed of a rather heterogenous group of Diptera-specific crystal protein genes. The *cry4A, cry4B, cry10A* and *cry11A* genes encode proteins with predicted molecular masses of 135, 128, 78 and 72 KDa respectively.

e) *cyt* genes: The 27 kDa protein encoded by *cyt1A* shows no sequence homology to the other crystal protein genes. It is cytolytic for a variety of invertebrate and vertebrate cells including mammalian erythrocytes. Cyt toxins are hemolytic and cytolytic *in vitro* and are specifically active against dipteran larvae *in vivo*. The proteins vary in size from 25 to 28 kDa and are smaller than Cry polypeptides. Three types have been defined according to immunoreactiving.

Cyt1A is the characteristic cytolytic toxin of *Bt* subsp. *israelensis* and *Bt* subsp. *morrisoni* PG-14. Cyt1Aa has been reported to be highly toxic to cottonwood leaf beetle (*Chrysomela scripta*) and its mode of action against *C. scripta* is similar to that observed in mosquito and blackfly larvae (Federici and Bauer, 1998). It is possible that Cyt proteins may have an even broader spectrum of activity against insects and owing to their different mechanism of action in comparison to Cry proteins, might be useful in managing resistance to Cry toxins.

A second Cyt2A protein is 2kDa larger than the Cyt1A protein and has been found in the inclusions of *Bt* subsp. *kyushuensis* and *Bt* subsp. *darmstadiensis* 73-E10-2. Cyt1A and Cyt2A show 70% functional homology (39% amino acid sequence identity) (Cannon, 1996 and Hofte and Whiteley, 1989). The CytC is representative of *Bt* subsp. *fukuskaensis*. A new type CytD, has been proposed for *Bt* subsp. *jegathesan* (Guerchicoff et al., 1997).

4.4   Structure of Bt *Crystal Proteins*

X-ray crystallography has revealed the three dimensional structure of the δ-endotoxin Cry3A from *Bacillus thuringiensis* subsp. *tenebrionis* (Li et al., 1991) that is specifically toxic to beetles (Figure 1.6), Cry1A (Grochulski et al., 1995) that is toxic to moth larvae and Cyt2A (Li, Koni and Ellar, 1996), a mosquitocidal toxin. Cry3A and Cry1Aa show about 36% amino acid sequence identity with each other. However, Cyt2A shows less than 20% amino acid sequence identity with Cry1Aa and Cry3A, and has a radically different structure.

Cry3A and Cry1Aa in contrast to Cyt2A, comprise of three domains; Domain I, is a helical bundle comprised of a centrally located hydrophobic α-helix surrounded by six additional amphipathic α-helices, thought to be pore-forming region. The Domain II, consists of eleven β-strands and one short α-helix, which are arranged as three β-sheets in a common 'Greek-key' motif. The β-sheets form the sides of a triangular structure enclosing a central hydrophobic core. The site directed mutagenesis experiments reveal that this domain is the one responsible for receptor binding. The Domain III of the *Bt* toxin consists of twelve additional β-strands in a 'jelly-roll' configuration. The role of the third

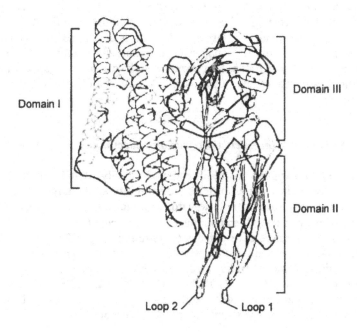

Figure 1.6   Line drawing of the three dimensional structure of Cry3 showing distinct structural domains I, II and III (Source: Li et al., 1991).

domain seems to be closely related to the receptor recognition specificity but its precise role is still unclear. The unique features of δ-endotoxins make them an interesting model protein for studying the relationship between structure and function (Powell et al., 1995; Yamamoto and Powell, 1993a and 1993b).

5   Insecticidal Activity

Since the first use of *Bt* as a biological control agent for insects, a great deal of progress has been made. New *Bt* strains with improved insecticidal activity have been identified. New strains with new activities have also been identified, thereby increasing the spectrum of activity of microbial sprays as a whole.

The host range of *Bt* has expanded considerably in recent years due to extensive screening programs, from species in three insect orders (Lepidoptera, Diptera, and Coleoptera) to species in eight insect orders (Homoptera, Orthoptera, Mallophaga, Hymenoptera, Siphanoptera) (Feitelson et al., 1992 and

Feitelson, 1993). In addition groups susceptible to *Bt* include certain protozoa, platyhelminths, nematodes, lice, aphids, mites, mollusks and flukes.

The insecticidal activity of a *Bt* strain is determined by (a) the number of cry genes present, (b) qualitative differences in amino acid sequences between ICPs, (c) difference in expression levels of the genes and (d) the intrinsic properties of the ICPs such as fermentation stability and activity (Cannon, 1996).

5.1   Variation in Insecticidal Activity and Specificity

The causes of variation in the insecticidal activity of ICPs include both the nature of the toxin and the intrinsic susceptibility of the target. The *Bt* strains differ not only in their activity towards specific insect species but also in the relative potency of their δ–endotoxins towards susceptible species. At least three factors influence the relative potency (or specificity) of ICPs: (a) the origin (i.e. the *Bt* strain) of the toxin, (b) the degree of solubility of the protoxin crystals in the gut juices of the host and (c) the intrinsic susceptibility of the host to the toxin (Jaquet et al., 1987). Thus the variability of the ICPs themselves, their solubility and the heterogeneity and concentration of receptors in the target organism contribute to the specificity of the larvicidal activity.

5.2   Determination of Insecticidal Activity

To determine the insecticidal activity of the crystal protein, the protein is fed to insects either mixed with an artificial diet or applied on the surface of leaf disks. Insect mortality is normally read after an incubation period of several days. Alternatively, feeding inhibition may be used as an indicator of the activity. These assays are examples of *in vivo*, bioassay methods. The crystal protein can also be assayed *in vitro*. In this case, crystal protein need to be activated by incubation with an appropriate proteinase. The activated toxin can then be assayed by either an insect cell lysis, toxin receptor-binding or patch clamp method (Yamamoto and Powell, 1993a).

Bioassay of a microbial insecticide and that of a chemical insecticide are conceptually different. In the case of a chemical, one determines the quantity and purity (quality) of the active ingredient and uses these data to judge the insecticidal activity of the product. On the other hand, microbial insecticides have no tangible active product to be measured. The level of impurities in the final product is very high; in fact, the active ingredient may represent only a small percentage of the final product. Furthermore, the level of impurities can vary widely from fermentation to fermentation.

In the case of bioinsecticides, one must measure the insecticidal activity of the product through bioassays and use this measurement to judge the quantity and quality of the product. Bioassays are not accessories to the production of bioinsecticides, they are central and vital to it and the monitoring of entire production process and the product itself is dependent upon these bioassays.

5.2.1 Bioassay of Microbial Insecticides

Bioassay requires determining the LC_{50} of the product against test insects. LC_{50}s by themselves can be misleading since the susceptibility of insect population to a particular microbial insecticide can vary from day to day. Therefore, it is preferable to compare the LC_{50} of each test sample with that of a standard formulation tested at the same time on the same population of insects. For convenience, a potency value is assigned to the standard in International Units (IUs) and the activity of the test sample is then also expressed in IUs compared to the standard.

Calculation of potency in International Units (I.U.) –

$$\text{Potency of Sample, IU/mg} = \frac{LC_{50} \text{ standard x Potency of standard}}{LC_{50} \text{ sample}}$$

The activity determined *in vivo* is greatly influenced by several factors, including diet, incubation time, incubation temperature, insect age, and insect behavior. Since the crystal protein is a stomach poison, a sufficient amount of the protein must be ingested to kill the insect. If the insect rejects the diet containing the crystal protein, the activity appears low. The rate of crystal solubilization may be another important factor that can influence activity. In addition, natural variation (a numerical difference in response that is detected each time a bioassay is repeated with one genetic strain, either within a single generation or >1 generation) in response during studies carried out to estimate potency of microbial pesticides, can lead to erroneous conclusions (Robertson et al., 1995). For these reasons, it is almost impossible to compare bioassay data reported from different researchers who use different bioassay methods.

5.2.2 Role of Spores in the Toxicity

The role of the spore in the toxicity of the δ-endotoxin has been subject to intense debate. Early studies on the insecticidal properties of *Bt* focussed on the spore, which at the time was more readily quantifiable than the δ-endotoxin components of fermentation products. The spore coat consists of proteins immunologically homologous with those of the crystal. Some insects, such as wax moth (*Galleria mellonella*) and the mediterranean flour moth (*Anagasta Kuehniella*) require both spores and crystals to be present to cause death. However, there are many species eg. *Pieris brassicae* and *Choristoneura fumiferana*, where spores play little or no role in mortality. Rossa and Mignone (1993) reported for *Bt* subsp. *israelensis* that a good spore count did not necessarily lead to high larvicidal potency. Milne et al. (1990) conclude that spore does have a role in determining insecticidal activity, particularly in the case of species generally less susceptible to *Bt*. In larvae of these species- eg. gypsy moth (*Lymantria dispar*), diamondback moth (*Plutella xylostella*) and the beet armyworm (*Spodoptera exigua*) - after an initial intoxication resulting from activation of the Cry toxins in the midgut, the spore germinates and produces enzymes (phospholipases and proteases). These contribute to the lysis of the gut cells by degrading cell membranes (Federici, 1998). Van Frankenhuyzen (1994) suggests that the bacterial specicemia makes an important contribution to mortality in *C. fumiferana* larvae, and can be initiated by a low level of spores in toxin-challenged larvae.

In terms of the commercialization of *Bt* formulations, the focus of attention has largely shifted away from spores, although there are reports that their presence can increase the toxicity of ICPs against certain species and also reduce the effect of *Bt* induced resistance (see Chapter 5).

5.2.3   Spore Count

It has been stated in the past that there is a direct and quantitative relationship between the growth of an insect pathogen and its toxicity toward insects, and therefore a bioassay test is not needed. In the case of *Bt*, it was suggested that a simple bacteriological plate count to measure the spores/ml of a fermentation broth or the spores/gm of a formulation will indicate how much insecticidal activity is present. Thus, spore counts could offer an attractive alternative to the bioassay. Unfortunately, spore counts are essentially meaningless in determining the potency of a microbial insecticide, as spore count does not reveal the quantity of δ-endotoxin produced by an isolate of *Bt*. Many attempts to replace the bioassay with spore counts have been made. These have however been unsuccessful.

6   Mode of Action

Bt has a wide range of well-characterized insecticidal crystal proteins, which express as protoxin during sporulation. The crystalline protoxins are inactive. To be insecticidal, these protoxin must first be ingested by the insect and proteolytically activated to form active toxin. This happens in the insect midgut, which is also the target organ for *Bt* toxin. Several recent reviews have considered the mechanisms or mode of action of Cry toxin (Gazit and Shai, 1995; Himeno and Ihara, 1995; Thompson et al., 1995; Wolfersberger, 1995; Knowles, 1994; Knowles and Dow, 1993; Yamamoto and Powell, 1993b and Gill et al., 1992).

6.1 Mechanism

The mechanism of action of insecticidal crystal protein of *Bt* is a multistep process. These include solubilization and processing of proteins, toxin binding, membrane interaction, pore formation, cell lysis and bacterial septicemia and death.

6.1.1 Solubilization and Processing of Protoxin

Crystal protein activation has been often referred to as a two step process, the first step involving liberation of the protoxin or crystal protein by dissolution of the crystal and the second step, proteolytic digestion of the protoxin yielding the active toxin. Thus, dissolution and digestion process has been termed as activation (Powell et al., 1995).

After ingestion, the crystal is first solubilized due to the extreme pH of the insect midgut, highly alkaline in Lepidoptera and highly acidic in Coleoptera larvae (Prieto-Samsonov et al., 1997). The alkaline-solubilized crystal proteins (usually soluble only above pH 9.5) from most *Bt* strains are about 130 KDa and require processing by insect gut proteinase. The action mediated by the alkaline pH and proteinases of insect midguts, yields 60 to 70 KDa proteinase resistant toxin fragments. The size of this active fragment varies with *Bt* strain.

The midgut environment can play a crucial role in specificity as shown with the activation of δ-endotoxin from *B. thuringiensis* subsp. *aizawai*. When activated with lepidopteran *Pieris brassicae* (European cabbage worm) midgut extract, the toxin kills both *P. brassicae* and dipteran *Aedes aegypti* (Yellow fever mosquito) larvae, however when activated by *Ae. aegypti* midgut extract, the isolate is toxic only to these mosquito larvae. Thus, depending on the proteolytic enzyme, a protoxin can be activated into either a dipteran or lepidopteran toxin (Koziel et al., 1993). Differences in the extent of solubilization sometimes explain differences in the degree of toxicity among Cry proteins (Aronson et al., 1991 and Du et al., 1994). A reduction in solubility is speculated to be one potential mechanism for insect resistance (McGaughey et al., 1992).

6.1.2   Toxin Binding

The activated toxins bind to specific proteins called receptors located on the apical brush border membrane of the columnar cells. Much of the host specificity of *Bt* toxins is dependent on toxin structure and presence of toxin receptor sites in the insect midgut. The Cry toxins bind to proteinaceous receptors in the cell membrane of the insect midgut epithelia. At least two groups of membrane receptors, cadherins and aminopeptidases have been identified. Cytolytic toxins, that have significantly different structure, appear to bind to membrane lipids (Gill et al., 1999). Binding, while essential is not sufficient to produce mortal damage and although receptor play an essential role, post-binding factors are required for successful intoxication by *Bt* δ-endotoxins (Bauer, 1995 and Peyronment et al., 1997).

6.1.3   Membrane Interaction

Upon binding to these specific receptors, the toxin then inserts into the cell membrane forming pores thereby disrupting cell function. Various modes were proposed to explain the role of toxin receptor in pore formation, however, the actual process at the molecular level is not well understood (Gill et al., 1992 and Knowles and Dow, 1993).

6.1.4   Pore Formation

The formation of toxin induced pores in the columnar cell apical membrane allow rapid fluxes of ions. Knowles and Ellar (1987) proposed the mechanism of Colloid-Osmotic lysis, which suggests the formation of small 0.5 –1 nm pores in the cell membrane of the epithelial cells, resulting in an influx of ions accompanied by an influx of water.

6.1.5   Cell Lysis

As a consequence of pore formation, the cells swell and eventually lyse. The model proposed by Knowles and Dow (1993) placed emphasis on the cessation of K+ pump that leads to the swelling of columnar cells and osmotic lysis. The disruption of gut integrity results in the death. The overall midgut pathology of *Bt* toxicity results in a loss of bassal involutions in the columnar cells; swelling of the apical microvilli vesticulation of the endoplasmic reticulum; loss of ribosomes; swelling of mitochondria; swelling of the cell and nucleus and subsequent rupture of nuclear, organelle and plasma membrane (figure 1.7) and

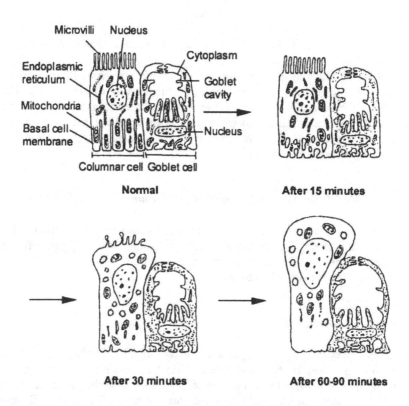

Normal

After 15 minutes

After 30 minutes

After 60-90 minutes

Figure 1.7 A schematic representation of the ultrastructural changes in the midgut epithelical cells of *Bombyx mori* induced by *Bacillus thuringiensis* δ-endotoxin. With lapsed time, in columnar cells – midgut epithelial cells show disappearance of microvilli and basal infolding, swollen nucleus with transformation of mitochondria into condensed form and the endoplasmic reticulum into vacuole-like configuration. Simultaneously, in goblet cells – high electron density of cytoplasm and enlargement of goblet cavity accompanied with enlargement of intra-space in basal infoldings take place (Source: Endo and Nishitsutsuji-Uno, 1980).

finally release of the cell contents into the lumen with sloughing of the cells. Other signs include increase in the number of size of nuclear pores, separation of cells from each other and from the basement membrane and nearly complete destruction of goblet cell (Endo and Nishitsutsuji-Uwo, 1980; Luthy and Ebersold, 1981; Lakhim-Tsor et al., 1983 and Bravo et al., 1992).

6.1.6   Bacterial Septicemia and Death

Heimpel and Angus (1959) provided the first concise account of mode of action of *Bt*. They noted that the insect midgut membranes are disrupted by the δ-endotoxin, allowing an ionic flow into the haemolymph. Death occurs when lysis of midgut cells cause irreparable breakdown of the midgut integrity, allowing *Bt* and other bacteria present in the lumen to gain access to the body cavity. The insect haemolymph provides an excellent medium for bacterial growth. Death caused by bacteria septicemia usually occurs within 2-3 days post-ingestion (Knowles, 1994).

6.2   Models for Mechanism of Pore Formation

There are various models proposed for mechanism of pore formation by *Bt* δ-endotoxin. Among the proposed models, the "umbrella" and "penknife" models, based on the Cry3A structure, have been reviewed by Knowles (1994). Both of these models are described below.

6.2.1   Umbrella Model

As described earlier, the Cry toxin structure consists of three protein domains each with a specific function. The domain I consist of seven α-helices and is thought to be involved with membrane interactions and the insertion of the toxin into the insect's midgut epithelium and pore formation. The domain II appears as a triangular column of three β-sheets and is reported to be involved in receptor binding. Domain III consists of a β-sandwich and is implicated in insect specificity and stability of the protein structure and might also participate in receptor binding (De Maggd et al., 1996).

The Umbrella Model for the mechanism of action of *Bt* δ-endotoxin (figure 1.8A) proposes that the first event in toxicity is the binding of domain II of the active toxin to an insect gut receptor. This binding triggers a conformational change of the protein, opening the toxin "Umbrella" and causing it to insert into the membrane forming deltaprotein lined pores that lead to cell lysis and the eventual death of the insect. In this model, helices α-4 and α-5 drop down into the membrane as a helical hairpin while the rest of the domain I flattens out. One possible orientation of the toxin as it binds to surface, is in an umbrella-like molten globule state (Figure 1.8B). It has been suggested that key feature of this model is that domain I helix α-5 might be the trans-membrane segment that formed the lining (Li et al., 1991).

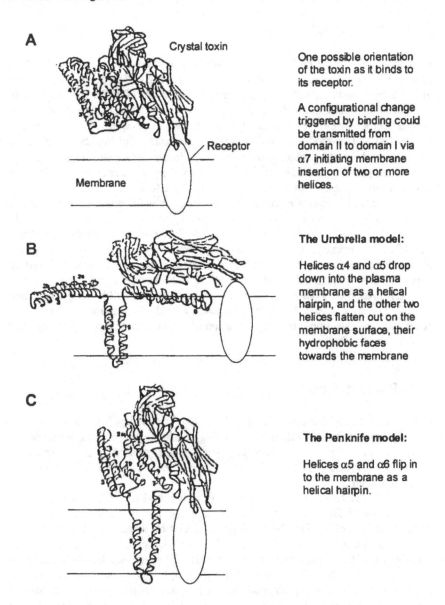

A

Crystal toxin

Receptor

Membrane

One possible orientation of the toxin as it binds to its receptor.

A configurational change triggered by binding could be transmitted from domain II to domain I via α7 initiating membrane insertion of two or more helices.

B

The Umbrella model:

Helices α4 and α5 drop down into the plasma membrane as a helical hairpin, and the other two helices flatten out on the membrane surface, their hydrophobic faces towards the membrane

C

The Penknife model:

Helices α5 and α6 flip in to the membrane as a helical hairpin.

Figure 1.8   Models of the Bt crystal toxin forming a pore through the insect cell membrane (Adapted from Knowles, 1994).

6.2.2   *Penknife Model*

Hodgman and Ellar (1990) proposed helices α-5 and α-6 as the pairs most likely to form the pore. Helices α-5 and α-6 are joined at the end of domain I predicted to be farthest away from the membrane and would therefore have to flip out of domain I like a penknife fashion and insert into the membrane. The remainder of the molecule would remain at the membrane surface or on the receptor (Figure1.8C). This model does not require rearrangement of the rest of domain I, although α-4 would probably have to slide downwards relative to α-3. The authors proposed a formation of hexameric toxin pore (internal pore radius 0.6 nm) lined by six helical hairpins, each donated by a toxin molecule.

Further studies are needed to provide conclusive proof for these or any other models that may be proposed to explain the mechanism of action of *Bt* toxins (Dean et al., 1996; Nunez-Valdez, 1997; Yamomoto and Powell 1993a and 1993b). Recently, a bivalent two-step binding model has been proposed which alternatively may represent a conformational change occurring. Based on gypsy moth receptor binding studies with a three-domain lepidopterin-specific toxin Cry1Ac, the model proposes a receptor recognition and binding to domain III followed by a slower but tighter secondary binding to domain II. Domain I potentially inserts upon binding to a receptor in a membrane environment (Jenkins et al., 2000).

7   Persistence

Bt has had considerable use in field and forest to control lepidopterous pests, but it rarely persists for more than a month to give any degree of long term control (Pruett et al., 1980). *Bt* spores can survive for several years after spray application (Addison, 1993), although rapid declines in population and toxicity have been noted.

The persistence of *Bt* preparations varies markedly according to habitat type. Density of *Bt*s in granaries is extremely high and probably correlates with the high density of insects, the stability of the climatological conditions, and the absence of sunlight (UV radiation), which break down ICPs. Only in such contained environments *Bt* spores and crystals are preserved for sufficient time to exceed the threshold dose needed for infection, killing larvae, and completing its growth cycle. The presence of *Bt* in a soil does not indicate any enhanced value in insect control. *Bt* rarely, if even can initiate an epizootic unless abetted by external conditions such as the crowding commonly present in insect rearing facilities (Lambert & Peferoen, 1992; Pruett et al., 1980 and Delucca et al., 1981).

7.1 Factors Affecting Persistence

The persistence of the introduced toxins is the function primarily of (a) the concentration added, (b) the rate of consumption and inactivation by insect larvae and (c) the rate of degradation by the microbiota (Tapp and Stotzky, 1995).

Two major technical problems are associated with the effective field use of microbials, their proper application i.e. their placement where and when they will exert the most control and their persistence i.e. keeping them active as long as the destructive stage of the pest population is present.

The loss of persistence of biological activity is the result of multiple environmental factors such as temperature, water and sunlight. The susceptibility of *Bt* to bio-degradation and inactivation under field conditions prevents its greater commercial uptake.

7.1.1 Temperature

In most agro-ecosystems ambient temperature during the growing season ranges from about 10 to 40°C; however, the optimum range for infection, growth and development for most entomopathogens lies between 10 and 30° C. In general, temperature with range of 10 to 30° C for less than 30 days (the period within which most crops would be vulnerable) does not effect the stability of many entomopathogens. However, deleterious effects can occur at temperatures less than 10°C or greater than 30° C, when entomopathogens are stressed by interaction with water, sunlight, foliage or soil chemicals etc. Temperature above 35° C generally inhibit growth and development of *Bt*. Insecticidal activity of *Bt* is markedly decreased as temperature approach 50°C (Table 1.3) (Ignoffo, 1992).

7.1.2 Water

Water, other than as a dispersal and diluting vehicle or in combinations with other environmental factors, may limit persistence and subsequent field effectiveness. The half life of *Bt* exposed at 50°C is greater than 100 days while wet spores have half-life of less than 60 days. The spores of *Bt* survive longer if they are dry. Estimated loss in stability (% in days) at 30°C for wet *Bt* spores is 18% while there is no loss in stability when it is dry (Ignoffo, 1992).

Table 1.3   *Bt* stability in days at various temperatures (Adopted from Ignoffo, 1992)

Inoculum	*5-10°C*	*20-30°C*	*45-50°C*
Spores	>5,000	>300	100
Toxin	>5,000	90	>90

7.1.3   pH

Soil pH may be an important variable affecting the survival of *Bt* in soil. Enumeration of *Bt* spores on nutrient agars of different pH showed that optimum growth occurred at pH 6.7 and 6.4. A 10-fold reduction in numbers occurred at6.0 and 5.6, a further 10,000-fold reduction at pH 5.1 and no growth at pH 4.4 (Saleh et al., 1970).

7.1.4   Solar Radiation

Natural sunlight (the active spectrum is between 290 and 400 nm) is the most destructive environmental factor affecting the persistence of entomopathogens. Sunlight may directly or indirectly inactivate entomopathogens. The direct effects may be deletions, cross-linking, strand breakage, and/or formation of labile sites on DNA. Indirect effects may be due to generation of highly reactive radicals (e.g. peroxides, hydroxyls, singlet oxygen) produced by near ultraviolet radiation (UV), which are primarily responsible for reducing the field persistence. If mechanical loss is excluded, the solar radiation would be major factor affecting the persistence of *Bt* on treated leaves (Salama et al., 1983 and Ignoffo, 1992).

The variability in reports on the persistence of *Bt* is probably a result of differences in type of foliage sprayed, the strain of *Bt* tested, weather conditions after the spray, differences in the relative susceptibility of the insects used in the bioassays and the extent of sunlight shielding provided by the plant.

8   Safety and Ecotoxicology of *Bt*

After more than three decades of operational, commercial applications of many types of *Bt* formulation on millions of acres of crops, forests, lakes, rivers and streams there has never been a reported, document incident involving adverse field effects on man or the environmental following these applications.

No unexpected toxicities have been noticed and no serous outbreak of *Bt* in insect populations has been documented. This is probably because, *Bt* does not survive or grow well in natural habitats, and has a narrow host specificity.

8.1   Mammalian Toxicity

There are concerns about the mammalian pathogenicity of the genus *Bacillus* because one member, *B. anthralis*, is a virulent mammalian pathogen, and various species of *Bacillus* have been associated with infections following traumatic wounds. For these reasons, *Bacillus thuringiensis* underwent careful testing during its commercial development, including full chemical safety tests and infectivity studies. Like their chemical counterparts, entomopathogens must be evaluated for their safety to both animals and humans, although the tests used to evaluate their safety differ from the chemical safety protocols. Microbial safety tests concentrate on acute toxicity and vertebrate infectivity, while chemical safety tests focus on acute toxicity, neurotoxicity, and carcinogenicity. The ocular and dermal irritancy of both chemical and microbial agents are additional concerns (Siegel and Shadduck, 1990).

There were no adverse effects on human volunteers, who were also fed *Bt*. Human ate 10^{10} spores per day for five days and inhaled 10^{9} spores with no ill effects (Burges, 1981). Injecting high dosages of *Bt* intracranially and intraoculary in domestic animals and wild type did not cause significant toxicity or infectivity. Conventional routes of exposure such as oral, parenteral, respiratory and dermal also showed no toxicity or pathogenicity (Saik et al., 1990).

The Cry proteins typically require both solubilization and activation steps before they become biologically active toxins. For most, solubilization occurs in the highly alkaline environment of lepidopteran insect midguts. Activation occurs via discrete proteolysis by insect gut enzymes and may occur concomitantly with the solubilization step. The highly acidic nature of most mammalian guts is not a favorable environment for the Cry toxin. The low pH of most mammal guts would solubilize and denature the Cry proteins, making them susceptible to hydrolysis by native gut proteases into inactive small peptides and free amino acids. As such, all the literature reports conclusively support the observation that *Bt* is not a mammalian pathogen.

8.2   Toxicity to Avian Species and Fish

Bt was fed to avian species for as long as 690 days with no adverse affect. *Bt* was found to be not toxic to wildlife including birds and fish and non-target vertebrates or invertebrates in the terrestrial or aquatic habitat (Burges, 1981).

8.3   Affect on Non-Target Organisms (NTOs)

One of the outstanding advantages of *Bt* is safety for insect parasites and predators. *Bt* has no apparent effect on beneficial insects such as honeybees (Bailey, 1971 and Burges, 1981). The lepidopteran active varieties of *Bt* however, exhibit mild toxic effects on some invertebrate NTOs at recommended label rates. Many of these effects are secondary in nature, resulting from the declining health of the host or prey larva (Melin and Cozi, 1990).

8.4   Affect on Beneficial Insects

8.4.1   Silkworm (Bombyx mori)

Concern over potential harm to silkworm has led some countries to prohibit the use of *Bt* product, a position that might now logically be resisted given the diversity of available *Bt* strain (Driesche and Bellow, 1996). *Bt* subsp. *thuringiensis* that are effective against pest insects, are relatively harmless to silkworms. However, what is harmless for silkworms is not necessarily harmless for other beneficial insects. e.g. *Bt* subsp. *thuringiensis* is more harmful than *Bt* subsp. *sotto* to honeybees, but subsp. *sotto* is one of the most harmful to silkworms (Bailey, 1971).

8.4.2   Honey-bees (Apis mellifera)

Research by several workers failed to reveal *Bt* activity against adult or immature honeybees. Bees were fed with sugar solution containing *Bt* subsp. *thuringiensis* spores (0.67 and 1.67 X 10^9 per bee), supernatant (2.5 mg per bee), and crystals (0.5 to 16×10^6 per bee) and also crystal of subsp. *alesti* and *sotto* (both at 0.5×10^6 per bee). All three crystal types failed to harm the bees, but the β-exotoxin (supernatant) gave nearly 100% mortality at 7th day. Significant mortality was seen in the spores treatment at 8th day. This was probably a consequence of septicemia (Cantwell et al., 1966).

9   Concluding Remarks

Because of their environmental safety, microbial insecticides are one of the few pesticides that can be developed and registered quickly and cheaply. Pressure for non-hazardous, environmentally compatible pest-control measures has spurred the interest in *Bt*, which is now widely acknowledged as an interesting and promising source of insecticides. In addition, resistance to conventional insecticides does not confer cross-resistance to *Bt* toxins due to the unique mode of action of δ-endotoxin. *Bt* toxins would contribute to reduction of the chemical insecticides load, reduced pressure on beneficial insects and other non-target organisms, and increase worker safety by reducing exposure to pesticides.

Ongoing screening programs will undoubtedly reveal *Bt*s with new activities in terms of increased toxicity and new spectra of activity. The highly specific toxicity of the ICPs has raised the interest of scientists and industry. As ICPs can be readily produced by fermentation, the development of relatively economical biopesticides is practical. The bacterial production of ICPs and its release as a crystal in a stable, inert form facilitates the production of commercial sprays.

The major impetus for greater use of *Bt* in agriculture is the development of resistance to conventional insecticides. Today, *Bt*-based insecticides are frequently used in intensive agriculture, (a) in conjunction with conventional insecticides (see Chapter 10) as a backup for control failure, or (b) last resort once resistance to other insecticides has occurred (Bauer, 1995).

Despite many appealing characteristics, the use of conventional *Bt*-based insecticides is often constrained due to (a) very narrow specificity, (b) maximum effectiveness being limited to particular developmental stage of the pest, (c) short shelf life, (d) low potency, (e) lack of systemic activity and (f) the presence of viable spores. The specificity is a problem for major world crops that have pest complexes and a single *Bt*-based product cannot control them. The insecticidal microorganisms or their toxic products are sensitive to environmental factors such as ultraviolet light, plant surface chemical, heat and desiccation. The *Bt* δ-endotoxin is short-lived on crops, necessitating the need for many applications during a growing season. *Bt*, as with other spray-on pesticides, is difficult to deliver to insect species that burrow into their host plant, hide under leaves or live primarily under the soil surface. In case of aquatic insect control, maintaining the δ-endotoxin in the water at the level of insect feeding zone is difficult. Because *Bt* kills insects slowly and has low residual activity, it must be used prophylactically, and *Bt*-based products cannot control well-established insect populations. The presence of spores in *Bt* products is a problem in countries like Japan, because a massive release of spores would potentially threaten the silk industry, which uses the *Bt*-sensitive silkworm.

Finally, potentially although *Bt* provides an alternative to chemical insecticides, totally substituting *Bt* for the use of chemicals would be a mistake. Use of the chemical arsenal in combination with *Bt* would probably enable a more judicious use of both and would also delay the onset of insect resistance.

References

1.   Addison, J. A., 1993. "Persistence and non-target effects of *Bt* in soil: a review", *Can. J. For. Res.*, 23, 2329-2342.

2.   Agaisse, H. and D. Lereclus, 1995. "How does *Bacillus thuringiensis* produce so much insecticidal crystal protein ?", *J. Bacteriol.*, 177(21), 6027-6032.

3.   Aronson, A. I., E-S. Han, W. McGaughey and D. Johnson, 1991. "The solubility of inclusion proteins from *Bacillus thuringiensis* is dependent upon protoxin composition and is a factor in toxicity to insects", *Appl. Environ. Microbiol.*, 57, 981-986.

4.   Bailey, L., 1971. "The safety of pest-insect pathogens for beneficial insects", in *Microbial control of insect and mites*, H. D. Burges and N. W. Hussey, eds., Academic Press, pp.491-505.

5.   Bauer, L. S., 1995. "Resistance: A threat to the insecticidal crystal protein of *Bacillus thuringiensis*", *Florida Entomol.*, 78(3), 414-443.

6.   Beegle, C. C. and T. Yamamoto, 1992. "History of *Bacillus thuringiensis* Berliner research and development". *Can. Entomol.*, 124, 587-616.

7.   Bernhard, K., P. Jarrett, M. Meadows, J. Butt, D. J. Ellis, G. M. Roberts, S. Pauli, P. Rodgers and H. D. Burges, 1997. "Natural isolates of *Bacillus thuringiensis*: worldwide distribution, characterization and activity against insect pests", *J. Invertebr. Pathol.*, 70, 59-68.

8.   Bernhard, K. and R. Utz, 1993. "Production of *Bacillus thuringiensis* insecticides for experimental and commercial uses", in *Bacillus thuringiensis, An Environmental Biopesticide: Theory and Practice*, P.F. Entwistle, J.S. Cory, M. J. Bailey and S. Higgs, eds., John Wiley & Sons, pp. 255-267.

9.   Bravo, A., S. Jansens and M. Peferoen, 1992. "Immunocytochemical localization of *Bacillus thuringiensis* insecticidal crystal proteins in intoxicated insects", *J. Invertebr. Pathol.*, 60, 237-246.

10.   Burges, H. D., 1981. "Safety, safety testing and quality control of microbial pesticides", in *Microbial control of Pests and Plant Diseases 1970-1980*, H. D. Burges, ed., Academic Press, NY., pp.738-767.

11.   Cambell, D. P., D. E. Dieball and J. M. Brackett., 1987. "Rapid HPLC assay for the δ-endotoxin of *Bacillus thuringiensis*", *J. Agri. Food Chem.*, 35(1), 156-158.

12.   Cannon, R. J. C., 1996. "*Bacillus thuringiensis* use in agriculture: A molecular perspective", *Biol. Rev.*, 71, 561-636.

13.   Cantwell, G. E., D.A. Knox, T. Lehnert and A. S. Michael, 1966. "Mortality of the honey bee, *Apis mellifera*, in colonies treated with certain biological insecticides", *J. Invertebr. Pathol.*, 8, 228.

14.   Chaufaux, J., M. Marchal, N. Gilois, I. Jehanno and C. Buisson., 1997. "Investigation of natural strains of *Bacillus thuringiensis* in different biotypes through out the world", *Can. J. Microbiol.*, 43, 337-343.

15.   Crickmore, N., D. R. Zeigler, J. Feitelson, F. Schnepf, B. Lambert, D. Lereclus, C. Gawron-Burke and D. H. Dean, 1995. "Revision of the nomenclature for *Bacillus thuringiensis cry* genes", *Program and Abstracts of the 28th Annual Meeting of the Society for Invertebrate Pathology*, Society for Invertebrate Pathology, Bethesda, MD, p. 14.

16. Crickmore, N., D. R.. Zeigler, J. Feitelson, E. Schnepf, J. Van Rie, D. Lereclus, J. Baum, and D. H. Dean, 1999. "Bacillus thuringiensis toxin nomenclature", http://www.biols.susx.ac.uk/Home/Neil_Crickmore/Bt/holo.html (site accessed on 23.5.1999)

17. de Barjac, H., 1981. "Identification of H-serotypes of *Bacillus thuringiensis*", in *Microbial Control of Pests and Plant Diseases*, (1970-1980), H. D. Burges, ed., Academic Press, USA, pp. 35-43.

18. de Barjac, H. and E. Frachon,1990. "Classification of *Bacillus thuringiensis* strains", *Entomophaga*, 35, 233-240.

19. De Maggd, R. A., M. S. G. Kwa, H. van der Klei, T. Yamamoto, B. Schipper, J. M. Vlak, W. J. Stiekema and D. Bosch, 1996. "Domain III substitution in *Bt* δ-endotoxin CryIA(b) results in superior toxicity for *Spodoptera exigua* and altered membrane protein recognition", *Appl Environ. Microbiol.*, 62(5), 1537-1543.

20. Dean, D. H., F. Rajamohan, M. K. Lee, S.-J. Wu, X. J. Chen, E. Alcantara and S. R. Hussain, 1996. "Probing the mechanism of action of *Bt* insecticidal proteins by site directed mutagenesis - a minireview", *Gene*, 179, 111-117.

21. Delucca, A. J., II, J. G. Simonson and A. D. Larson, 1981. "*Bacillus thuringiensis* distribution in soils of the United States", *Can. J. Microbiol.*, 27, 865-870.

22. Driesche, R. G. V. and Bellow, T. S., Jr., 1996. "Augmentation: Pathogen and Nematodes", in *Biological Control*, Chapman and Hall, pp. 229-231.

23. Du, C., P. A. W. Martin and K. W. Nickerson, 1994. "Comparison of disulfide contents and solubility at alkaline pH of insecticidal and non-insecticidal *Bacillus thuringiensis* protein crystals", *Appl. Environ. Microbiol.*, 60, 3847-3853.

24. Dulmage, H. T., 1983. "Guidelines for production of *Bacillus thuringiensis* H-14", *Proceedings of Consultation*, UNDP/WHO special programme for research and training in tropical diseases, 25-28 Oct. 1982., M. Vandekar and H.T. Dulmage, eds., Geneva, Switzerland, pp. 124.

25. Dulmage, H. T., 1970. "Insecticidal activity of HD-1, a new isolate of *Bacillus thuringiensis* var. *alesti*", *J. Invertebr. Pathol.*, 15, 232-239.

26. Dulmage, H. T. and R. A. Rhodes, 1971. "Production of pathogens in artificial media", in *Microbial Control of Insect and Mites*, H. D. Burges and N. W. Hussey, eds., Academic Press, pp.507-540.

27. Endo, Y. and J. Nishitsutsuji-Uwo, 1980. "Mode of action of *Bacillus thuringiensis* δ-endotoxin: Histopathological changes in the silkworm midgut", *J. Invertbr. Pathol.*, 36, 90-103.

28. Estruch, J. J., G. W. Warren, M. A. Mullins, G. J. Nye, J. A. Craig and M. G. Koziel, 1996. "Vip 3A, a novel *Bacillus thuringiensis* vegetative insecticidal protein with a wide spectrum of activities against lepidopteran insects", *Proc. Natl. Acad. Sci. USA*, 93, 5389-5394.

29. Federici, B. A., P. Luthy and J.E. Ibarra, 1990. "The parasporal body of *Bti*: Structure, protein composition and toxicity", in *Bacterial Control of mosquitoes and blackflies; Biochemistry, Genetics, and Application of Bti and BS*, H. de Barjac, D. J. Sutherland, eds., Unwin Hyman, London, pp. 16-44.

30.   Federici, B. A., 1993. "Insecticidal bacterial protein identify the midgut epithelium as a source of novel target sites for insect control", *Arch. Insect Biochem. Physiol.*, 22, 357-371.

31.   Federici, B. A., and L. S. Bauer, 1998. "Cyt1Aa protein of *Bacillus thuringiensis* is toxic to the cottonwood leaf beetle, *Chrysomela scripta*, and suppresses high levels of resistance to Cry3Aa", *Appl. Environ. Microbiol.*, 64(11), 4368-71.

32.   Federici, B. A., 1998. "Broadscale use of pest-killing plants to be true test", *California Agriculture*, 52(6), 14-20.

33.   Feitelson, J. S., J. Payne and L. Kim, 1992. "*Bacillus thuringiensis:*insect and beyond", *Bio/Technol.*, 10, 271-275.

34.   Feitelson, J. S., 1993. "The *Bacillus thuringiensis* family tree", in *Advanced Engineered Pesticides*, L. Kim, ed., Marcel Dekker, Inc., New York, pp. 63-71.

35.   Gazit, E., and Y. Shai, 1995. "The assembly and organization of the α-5 and α-7 helices from the pore – forming domain of *Bacillus thuringiensis* δ-endotoxin", *J. Biol. Chem.*, 270, 2571-2578.

36.   Gill, S. S., H. K. Lee, D. I. Oltean, M. Zhuang, M. D. Kawalek, and H. Cheong, 1999. "Structure and mode of action of *Bacillus thuringiensis*", Abstracts, American Chemical Society Meeting, March, 1999, 88.

37.   Gill, S. S., E.A. Cowles and P. V. Pietrantonio , 1992. "The mode of action of *Bacillus thuringiensis* endotoxins". *Annu. Rev. Entomol.*, 37, 615-636.

38.   Goldberg, L. J. and J. Margalit, 1977. "A bacterial spore demonstrating rapid larvicidal activity against *Anopheles sergentii, Uranotaenia unguiculata, Culex univitattus, Aedes aegypti and Culex pipiens*", *Mosq. News*, 37, 355-358.

39.   Gonzalez, J. M., Jr., and B. C. Carlton, 1984. "A large transmissible plasmid is required for crystal toxin production in *Bacillus thuringiensis* var. *israelensis*", *Plasmid*, 11, 28-38.

40.   Grochulski, P., L. Masson, S. Borisova, M. Pusztai-Carey, J. L. Schwartz, R. Brousseau and M. Cygler, 1995. "*Bacillus thuringiensis* Cry1A(a) insecticidal toxin: crystal structure and channel formation", *J. Mol.. Biol.*, 254, 447-464.

41.   Guerchicoff, A., R. A. Ugalde and C. P. Rubinstein, 1997. "Identification and characterization of a previously undescribed *cyt* gene in *Bacillus thuringiensis* subsp. *israelensis*", *Appl. Environ. Microbiol.*, 63(7), 2716-2721.

42.   Heimpel, A. M. and T. A. Angus, 1959, "The site of action of crystalliferous bacteria in Lepidoptera larvae", *J. Insect Pathol.*, 1, 152-170.

43.   Himeno, M., and H. Ihara, 1995. "Mode of action of δ-endotoxin from *Bacillus thuringiensis* var. *aizawai*", in *Molecular Action of Insecticides on Ion Channels*, J. M. Clark, ed., American Chemical Society, Washington, DC.

44.   Hodgman, T. C., and D. J. Ellar, 1990. "Models for the structure and function of the *Bacillus thuringiensis* δ-endotoxins determined by compilational analysis", *DNA Seq.- J. DNA Seq.Mapp.*, 1, 97-106.

45.   Hofte, H. and H. R. Whiteley, 1989. "Insecticidal crystal proteins of *Bacillus thuringiensis*", *Microbiol. Rev., 53(2), 242-255.*

46.   Ignoffo, C. M., 1992. "Environmental factors affecting persistence of Entomopathogens", *Florida Entomol.*, 75(4), 516-525.

47.   Itoua-Apoyolo, C., L. Drif, J. M. Vassal H. de Barjac, J. P. Bossy, F. Leclant and R. Frutos, 1995. "Isolation of multiple subsp. of *Bacillus thuringiensis*

from a population of European sunflower moth, *Homoeosoma nebulella*", *Appl and Environ. Microbiol.*, 61(12), 4343-4347.

48. Jaquet F., R. Hutter and P. Luthy, 1987. "Specificity of *Bacillus thuringiensis* delta-endotoxin", *Appl. Environ. Microbiol.*, 53(3), 500-504.

49. Jenkins, J. L., M. K. Lee, A. P. Valaitis, A. Curtiss, and D. H. Dean, 2000. "Bivalent sequential binding model of a *Bacillus thuringiensis* toxin to gypsy moth aminopeptidase N receptor", *J. Biol. Chem.*, 275, 14423-14431.

50. Keller, B. and G. A. Langenbruch, 1993. "Control of coleopteran pests by *Bacillus thuringiensis*", in *Bacillus thuringiensis, An Environmental Biopesticide: Theory and Practice*, P.F. Entwistle, J. S. Cory, M. J. Bailey and S. Higgs, eds., John Wiley & Sons Ltd., pp. 171-191.

51. Knowles, B. H. and D. J. Ellar, 1987. "Colloid-osmotic lysis is a general feature of the mechanism of action of *Bacillus thuringiensis* δ-endotoxins with different insect specificities", *Biochim. Biophys. Acta,*, 924, 509-518.

52. Knowles, B. H. and J. A. T. Dow, 1993. "The crystal delta-endotoxins of *Bacillus thuringiensis:* models for their mechanism of action on the insect gut", *Bioassays*, 15, 469-476.

53. Knowles, B. H., 1994. "Mechanism of action of *Bacillus thuringiensis* insecticidal δ-endotoxin", *Adv. Insect Physiol.*, 24, 275-308.

54. Koziel, M. G., N. B. Carozzi, T. C. Currier, G. W. Warren and S. V. Evola, 1993. "The insecticidal crystal proteins of *Bacillus thuringiensis*: Past, Present and future uses", *Biotech. Genet. Eng. Rev.*, 11,171-228.

55. Krieg, A., W. Schnetter, A. M. Huger and G.-A. Langenbruch, 1987. "*Bacillus thuringiensis* subsp. *tenebrionis* strain BI 256-82: a third pathotype within the H-serotype 8a8b", *System. Appl. Microbiol.*, 9, 138-141.

56. Krieg, A., A. M. Huger, G. A. Langenbruch and W. Schnetter, 1983. "*Bacillus thuringiensis* var. *tenebrionis*: a new pathotype effective against larvae of Coleoptera", *J. Appl. Entomol.*, 96, 500-508.

57. Kumar, P. A., R. P. Sharma and V. S. Malik, 1996. "The Insecticidal protein of *Bacillus thuringiensis*", *Adv. Appl. Microbiol.*, 12, 1-43.

58. Lakhim-Tsror, L., C. Pascar-Gluzman, J. Margalit and Z. Barak, 1983. "Larvicidal activity of *Bacillus thuringiensis* subsp. *israelensis*, serovar H-14 in *Aedes aegypti*: Histopathological studies", *J. Invetebr. Pathol.*, 41, 104-116.

59. Lambert, B. and M. Peferoen, 1992. "Insecticidal promise of *Bacillus thuringiensis*. Facts and mysteries about a successful biopesticide", *Bioscience*, 42(2), 112-122.

60. Li, J., J. Carroll and D. J. Ellar, 1991. "Crystal structure of insecticidal δ-endotoxin from *Bacillus thuringiensis* at 2.5 A° resolution", *Nature*, 353, 815-821.

61. Li, J., P. A. Koni and D. J. Ellar, 1996. "Structure of the mosquitocidal δ-endotoxin CytB from *Bacillus thuringiensis* subsp. *kyushuensis* and implications for membrane pore formation", *J. Mol. Biol.*, 257, 129-152.

62. Lopez-Meza, J. and J. E.Ibarra, 1996. "Characterization of a novel strain of *Bacillus thuringiensis*", *Appl. Environ. Microbiol.*, 62(4), 1306-1310.

63. Lopez-Meza, J. E., B. A. Federici, J. J. Johnson and J. E. Ibarra, 1995. "Parasporal body from *Bacillus thuringiensis* subsp. *kenyae* composed of a novel composition of inclusions and Cry proteins", *FEMS Microbiol. Letts.*, 134, 195-201.

64. Luthy, P. and H. R. Ebersold, 1981. *"Bacillus thuringiensis* δ-endotoxin: Histopathology and molecular mode of action", in *Pathogensis of Invertebrate Microbial Diseases*, E.W. Davidson, ed., Allenheld, Osmun and Co., Totowa, N.J., pp. 235-267.

65. Masson, L., M. Bosse, G. Prefontaine, L. Peloquin, P. C. K. Lau and R. Brousseau, 1990. "Characterization of parasporal crystal toxins of *Bt* subsp. *kurstaki* strains HD-1, and NRD-12", in *Analytical chemistry of Bacillus thuringiensis*, L.A. Hickle and W.L. Finch, eds., American Chemical Society, Washington, D.C., pp. 61-69.

66. McGaughey, W. H., and M. E. Whalon, 1992. "Managing insect resistance to *Bacillus thuringiensis* toxins", *Science*, 258, 1451-1455.

67. Melin, B. E. and E. M. Cozzi, 1990. "Safety to non-target invertebrates of lepidopteran strains of *Bacillus thuringiensis* and their δ-endotoxins", in *Safety of Microbial Insecticides*, M. Laird, L. A. Lacey and E. W. Davidson, eds., CRC Press Inc., Florida, pp.149-166.

68. Mike, A., M. Ohba and K. Aizawa, 1990. "Extra-cellular production of a heat-stable somatic antigen by *Bacillus thuringiensis*", *Lett. Appl. Microbiol.*, 11, 247-250.

69. Milne, R., A. Z. Ge, D. Rivers and D. H. Dean, 1990. "Specificity of insecticidal crystal proteins: Implications for industrial standardization", in *Analytical Chemistry of* Bacillus thuringiensis, L. A. Hickle and W. L. French, eds., American Chemical Society, Washington, DC, pp. 22-35.

70. Nunez-Valdez., E. 1997. *"Bacillus thuringiensis* conference in Thailand: A widening "umbrella", *Nature Biotechnol.*, 15, 225-226.

71. Perferoen, M., 1992. "Engineering of insect-resistant plants with *Bacillus thuringiensis* crystal protein genes", in *Plant genetic manipulation for crop production*, A. M. R. Gatehouse, V. A. Hilder and D. Boulter, eds., Biotechnology in Agriculture, CAB Int'l., 7, pp. 135-153.

72. Peyronment, O., V. Vachon, R. Brousseau, D. Baines, J-L Schwartz and R. Laprade, 1997. "Effect of *Bacillus thuringiensis* toxins on the membrane potential of lepidopteran insect midgut cells", *Appl. Environ. Microbiol.*, 63(5), 1679-1684.

73. Powell, G. K., C. A. Charlton and T. Yamamoto, 1995. "Recent Advances in structure and function research on *Bacillus thuringiensis* crystal proteins", in *Bacillus thuringiensis Biotechnology and Environmental Benefits*, Hua Shiang Yuan Publ. Co. Taiwan, I, pp. 1-20.

74. Prieto-Samsonov, D. L., R. I. Vazquez-Parelan, C. Ayra-Pardo, J. Gonzalez-Cabrera and G. A. de la Riva, 1997. *"Bacillus thuringiensis*: From biodiversity to biotechnology", *J. Industrial Microbiol. Biotechnol.*, 19, 202-219.

75. Pruett, C. J. H., H. D. Burges and C. H. Wyborn, 1980. "Effect of exposure to soil on potency and spore viability of *Bacillus thuringiensis*", *J. Invertebr. Pathol.*, 35, 168-174.

76. Robertson, J. L., H. K. Preisler, S. S. Ng, L. A. Hickel and W. D. Gelernter, 1995. "Natural variation: a complicating factor in bioassays with chemical and microbial pesticides", *J. Econ. Entomol.*, 88(1), 1-10.

77. Rossa, C. and C. Mignone, 1993. "Delta-endotoxin activity and spore production in fed batch cultures of *Bt*", *Biotechnol. Letts.*, 15, 295-300.

78. Rowe, G. E. and A. Margaritis, 1987. "Bioprocess developments in the production of bioinsecticides by *Bacillus thuringiensis*", *Crit. Rev. Biotechnol.*, 6, 87-127.

79. Saik, J. E., L. A. Lacey and C. M. Lacey, 1990. "Safety of microbial insecticides to vertebrates - Domestic animals and Wildlife", in *Safety of Microbial Insecticides*, L. Marshall, L. A. Lacey and E. W. Davidson, eds., CRC Press, Florida.

80. Salama, H., M. S. Foda, F. N. Zaki and A. Khalafallah, 1983. "Persistence of *Bacillus thuringiensis* Berliner spores in cotton cultivations", *Z. Angew. Entomol.*, 95, 321-326.

81. Saleh, S. M., R. F. Harris and O. N. Allen, 1970. "Fate of *Bacillus thuringiensis* in soil: Effect of soil pH and organic amendment", *Can .J. Microbiol.*, 16, 677-680.

82. Schnepf, E., N. Crickmore, J. Van Rie, D. Lereclus, J. Baum, J, Feitelson, D. R. Zeigler and D. H. Dean, 1998. "*Bacillus thuringiensis* and its pesticidal crystal proteins", *Microbiol. Molecular Biol. Revs.*, 775-806.

83. Siegel, J. P. and J. A. Shadduck, 1990. "Mammalian safety of *Bacillus thuringiensis israelensis*", in *Bacterial Control of Mosquitoes and Blackflies*, H. de Barjac and D. J. Sutherland, eds., Unwin Hyman, London, pp. 202-217.

84. Smith, G. P., J. D. Merrick, E. J. Bone and D. J. Ellar, 1996. "Mosquitocidal activity of the CryIC δ-endotoxin from *Bacillus thuringiensis aizawai*", *Appl. Environ. Microbiol.*, 62(2), 680-684.

85. Sonngay, S. and W. Panbangred, 1997. As reported in Schnepf et al., 1998.

86. Stabb, E. V., L. M. Jackson and J. Handelsman, 1994. "Zwittermicin A-producing strain of *Bacillus cereus* from diverse soils", *Appl. Environ. Microbiol.*, 60, 4404-4412.

87. Steinhaus, E. A., 1951. "Possible use of *Bacillus thuringiensis* as an aid in the biological control of the alfaalfa caterpillar", *Hilgardiea*, 20, 359-381.

88. Tapp, H. and G. Stotzky, 1995. "Dot blot ELISA for monitoring the fate of insecticidal toxins from *Bacillus thuringiensis* in soil", *Appl. Environ. Microbiol.*, 61(2), 602-609.

89. Thompson, M. A., H. E. Schnepf, and J. S. Feitelson, 1995. "Structure, function and engineering of *Bacillus thuringiensis* toxins", in *Genetic Engineering: Principles and Methods*, Vol. 17, J. K. Setlow, ed., Plenum Press, New York, NY, pp. 99-117.

90. Tyagi, R. D., 1997. "Raw materials for the production of biopesticides in developing countries", *TCDC Intl. Workshop on Application of Biotechnology in Biofertilizers and Biopesticides*, Oct.15-18, 1997, IIT Delhi, Abstracts, pp.89-92.

91. van Frankenhuyzen, K., 1994. "Effect of temperature and exposure time on toxicity of *Bacillus thuringiensis* Berliner spray deposits to spruce budworm, *Choristoneura fumiferana* clemens (Lepidoptera: Tortricidae)", *Can. Entomol.*, 122, 69-75.

92. Villafana-Rojas, J., E. Gutierrez and M. de la Torre, 1996. "Primary separation of the entomopathogenic products of *Bacillus thuringiensis*", *Biotechnol. Prog.*, 12, 564-566.

93.   Wolfersberger, M. G., 1995. "Permeability of *Bacillus thuringiensis* Cry I toxin channels", in *Molecular Action of Insecticides on Ion Channels*, J. M. Clark, ed., American Chemical Society, Washington, DC.

94.   Yamamoto, T. and G. K. Powell, 1993a. "Structure and function of the insecticidal protein produced by *Bacillus thuringiensis*", in *Recent Adv. Mol. Biochem. Res. Proteins*, Proc., IUBMB Symp. Struct. Funct. 1992 (Publ. 93). Yau-huei Wei, Ching-San Chen and Jong-Ching , eds., World Sci., Singapore, pp. 137-144.

95.   Yamamoto, T. and G. K. Powell, 1993b. "*Bacillus thuringiensis* crystal proteins: Recent advances in understanding its insecticidal activity", In *Advanced Engineered Pesticides*, Leo Kim, ed., Marcel Dekker, Inc., New York. NY, pp. 3-42.

2

Bacterial Insecticides for Crop and Forest Protection and Insect Vector Control

1   Introduction

Microorganism based biopesticides have attracted greater attention during last decades for biological control of plant pests. The use of microorganisms as a source of biological compounds for insect pest control started with the discovery of the highly insecticidal bacteria *Bacillus thuringiensis*.   By far, the most successful of microbial insecticide in agricultural and forest insect pest control is *Bacillus thuringiensis* subsp. *kurstaki (Btk)*. Until the mid 1970s, *Bt* (mainly *Btk*) was thought only to produce proteins that were insecticidal to lepidopterous insects. In 1976, Goldberg and Margalit isolated a new subspecies of *Bt* named *israelensis (Bti)* which was found to be a pathogen of mosquito larvae. This discovery of a new subspecies of *Bt* led to much research into its possible use of mosquito control. After the successful development of *Bti* as a larvicide for some mosquitoes and blackflies, another *Bacillus* species, *sphaericus* was developed to complement *Bti* in the control of various mosquito species in diverse habitats. Detailed descriptions of individual characteristics, insecticidal properties and safety profiles of all the three above mentioned *Bacillus* species are provided in this chapter.

2   *Bacillus thuringiensis* subsp. *kurstaki*

Btk was first isolated in 1962 by Edourad Kurstak from diseased Mediterrean flour moth (*Anagasta kuehniella*) larvae from a flour mill at Bures Sur Yvette near Paris, France. Abbott Labs introduced the first commercial Product 'DiPel' in 1970. Since then, many isolates of *Btk* eg. HD-1 isolated from diseased pink bollworm (*Pectinophora gossypiella*) by Dulmage (1970) and HD-263 isolated from dead almond moth (*Ephestia cautella*) by Ayerst proved to be superior against all *Heliothis* spp. tested.

   Btk specificity for Lepidoptera and short persistence in the environment makes it an attractive alternative to synthetic chemical insecticides in many agricultural and forest ecosystems. It is widely used against lepidopteran pests in cotton, corn and soybean crops and is effective against foliage feeding caterpillars. It has also been successful in controlling forest pests infestations such as spruce budworm and gypsymoth (Spear, 1987).

   There are also some other lepidopteran-active *Bt* subspecies. These include *aizawai* HD-112 and HD-133 and *morrisoni* HD-12. Also, there is a coleopteran-specific *Bt* subsp. *tenebrionis (sandiego)*, which is effective against several beetle species such as, Colorado potato beetle (*Leptinotarsa decemlineta*) Alder leaf beetle (*Agelastica alni*), Cottonwood leaf beetle (*Chrysomela scripta*), Eucalyptus tortoise beetle (*Paropsis charybdis*) etc. (Keller and Langenbruch, 1993).

2.1   *Characteristics*

   Btk is widely distributed, rod shaped, aerobic, gram-positive, crystalliferous, spore forming, soil bacteria belonging to the family Bacillaceae. *Btk* is given a serotype name 3a3b. It is differentiated from other *Bt*s by comparing their serotypes.

   During sporulation, in addition to endospore, it produces a parasporal body that contains one or more proteins typically in a crystalline form. These crystalline proteins known as δ-endotoxins occur as protoxins, which upon activation exhibit a highly specific insecticidal activity against lepidopterous larvae.

2.2   Btk *Crystal Proteins and Genes*

2.2.1   *Parasporal crystal*

Most isolates of *Btk* (e.g. HD-73) produce a large single bipyramidal parasporal crystal (1.1 x 0.5µm) containing single protein that is almost always only toxic to lepidopterous insects. However in some isolates of *Btk* (e.g. HD-1), the bipyramidal crystal is accompanied by smaller cuboidal crystal and are toxic to lepidopterans and mosquitoes (Figure 2.1) (Federici, 1993; Cannon, 1996).

The bipyramidal protein crystal has a composition as follows: Cry1Aa, 13.6%, Cry1Ab, 54.2% and Cry1Ac, 32.2% (Masson et al., 1990). On the other hand, the cuboidal crystal inclusion consists of Cry2A proteins, which is responsible for weak but significant toxicity to mosquito larvae in some *Btk* strains such as HD-1, HD-263 and NRD-12. The *kurstaki* strain HD-263 showed superior activity when compared to HD-1 against several major agricultural pests although the insecticidal spectrum of the latter is broader. HD-73 is more active than HD-1 against *Heliothis* spp. and also Codling moth (*Cydia pomonella*) (Navon, 1993).

Btk HD-1 protoxin genes show 99% homology with *Bt* subsp. *sotto*, 91% with *berliner* and 85% with *kustaki* HD-73. The bipyramidal protein crystals of both *Btk* strains, NRD-12 and HD-1, contain all the three *cry1A* gene products (*cry1Aa*, *-b* and *-c*); however only the Cry1Aa and Cry1Ab component vary between crystals (Masson et al., 1990).

2.2.2   Btk *encoding genes*

cry1A gene appears to be the most widely distributed gene amongst different *Btk* strains and encode 130-160 kDa protoxins. These Lepidoptera specific proteins are converted by proteolysis into a toxic core fragment of 60 to 70 kDa (Hofte and Whiteley, 1989). Similarly, the genes of the class *cry2* that encode 65 kDa proteins also maintain Lepidoptera activity and occur in several *Btk* strains - such as HD 1, HD-263 and NRD-12. *cry2A* gene product is active against both

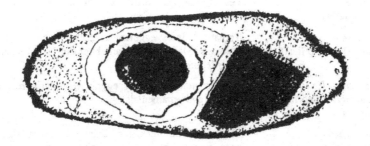

Figure 2.1 Electron micrograph of *Bacillus thuringiensis* subsp. *kurstaki* strain HD-1 showing crystal and spore (Spear, 1987).

Lepidoptera and Diptera species. Of the five protoxin genes occurring in *Btk* HD-1, four of its *cry* genes (*cryIAa, cryIAc, cry2Aa* and *cry2Bb)* are carried on a single 110-MDa plasmid (Carlton and Gonzalez, 1985) and the remaining *cry* gene (*cryIAb*) occurs on an unstable, smaller 44-MDa plasmid. The *cry* gene from *Btk* strain HD-1 was first cloned by recombinant DNA techniques in 1981. Since then many protoxin genes have been cloned and sequenced (Cannon, 1996).

2.3   Insecticidal Activity

Btk acts specifically against many species of Lepidoptera, with the most notable being the cabbage looper, tobacco hornworm, tobacco budworm, Europeon corn borer, gypsy moth and spruce budworm. Uses for which *Btk* is an accepted insecticide range widely and include forestry, vegetables, corn, tobacco, ornamentals, fruit trees and stored grains. The success of *Btk* is based on a combination of efficacy and safety.   Several pests of agronomic importance controlled by *Btk* and the related target crops are given in Table 2.1.

Table 2.1   Common pests controlled by *Btk* and their target crops

Pests	Target crops
Anticarsia gemmatalis (Velvetbean caterpillar)	Soybean, sunflower and peanuts.
Argyrotaenia spp. (Tortrix moth)	Pomefruits, currants (blueberries, caneberries, black-berries, dewberries, raspberries, strawberries), citrus (orange, lemon, grapefruits, tangerine, pamelo).
Choristoneura spp. (Spruce budworm)	Peanuts, soybean, forestry and shade trees.
Estigmene acrea (Saltmarsh caterpillar)	Tomato, peppers, eggplant, safflower, sugarbeet, mints, grapes, small fruits, cane and bush berries, soybean and other legume crops.
Helicoverpa/ Heliothis spp. (Budworm/fruitworm bollworms)	Tobacco, tomato, pepper, eggplant, corn/ maize, cotton, flowers and ornamentals.
Hyphantria spp. (Fall webworm)	Stone fruits, nut trees, pomegranates, forestry, shade trees, sugar maple trees ornamentals and flowers.
Lobesia botrana (Grape moth)	Grapes

Pests	Target crops
Lymantria dispar (Gypsy moth)	Pome fruits, forestry and shade trees, sugar maple trees, flowers and ornamentals.
Malacosoma spp. (Tent caterpillar)	Pecan, walnut, pome fruits, stone fruits,(almond, cherries, peach, nectarines), forestry and shade trees.
Mamestra brassicae (Cabbage moth)	Cabbage, cauliflower, brassicas, broccoli, brussels sprout, collard, kale, mustard, leafy vegetables (celery, cicory, saled), and sugarbeets.
Manduca spp. (Hornworm)	Cole crops and vegetables, tobacco, tomato, pepper and eggplant, flowers and ornamentals
Ostrinia nubilasis (European cornborer)	Corn/maize and other cereals.
Phologophora metuculosa (Angleshade moth)	Leafy vegetables, (celery, cicory, saled) tomato, pepper, and eggplant.
Platiphena scabra (Green cloverworm)	Peanuts, soybean, pasture (alfalfa, clover, grasses), cole crops and vegetables, potatoes, cucurbits and sunflower.
Platinota stultana (Omnivorous leafroller)	Flowers and ornamentals, grapes, small fruits, cane and bush berries, stone fruits, nut trees and pomegranates.
Plutella xylostella (Diamondback moth)	Flowers and ornamentals, cabbage, cauliflower, brassicas, broccoli, brussels sprout collard, kale and mustard.
Spodoptera spp. (Armyworms)	Cotton, sugar beet, tobacco, corn/maize, grapes, small fruits, berry, hops, alfalfa,,cole crops and vegetables, soybean and other legume crops, cucurbits, tomatoes, pepper and eggplant.
Trichoplusia ni (Cabbage looper)	Cabbage, cauliflower, brassicas, broccoli, brussels sprout collard, kale, mustard, beans, peas, tomato, pepper, eggplant, cucurbits (melon, water melon, cantaloupe, squash, cucumber), sugarbeet, mint, tobacco, cotton and alfalfa.
Tortrix sp. (Leaffolders)	Flowers and ornamentals, forestry and shade trees.
Udeaa ferrygallis (Leaf trier)	Leafy vegetables (celery, cicory, saled), peanuts, soybean, flowers and ornamentals.

Various lepidopteran-specific commercial products based on different strains of *Bt-kurstaki* are given in Table 2.2.

Table 2.2   Commercial products based on different strains of *Bt* subsp. *kurstaki*

Btk strain	Crystal Protein Composition	Product Trade Name	Manufacturer
HD-1	Cryl Aa, Cryl Ab, Cryl Ac, Cry2Aa, Cry2Ab	DiPel Thuricide	*Abbott Labs. Thermo Trilogy
NRD-12	Cryl Aa, Cryl Ab, Cryl Ac, Cry2A	Javelin	Thermo Trilogy
HD-263	Cryl Ab, Cryl Ac, Cry2A	BMP-123	Becker Microbial
SA-11		Delfin	Thermo Trilogy

* Since February 2000, Abbott Ag specialities products are owned by Valent BioSciences Corporation, a subsidiary of Sumitomo Chemical Company.

2.4   Mode of Action

Mode of action of *Bt* has been described in Section 6.0 of Chapter 1. *Btk* is a stomach poison and has no contact action. Larvae stop feeding on the treated plants within short time after the ingestion of a lethal dose of *Btk*. Resulting death usually occur within 2 to 3 days without further feeding. In commercial terms, a larva that stops feeding is no longer considered a pest. Only larvae are susceptible, whereas eggs or adults are not affected. Surviving larvae are particularly susceptible to natural controlling forces such as insect viruses, fungi, parasites, predators and environmental stress from weather extremes. Even larvae surviving to pupation may give rise to pupae below average weight and adults, which may be small, deformed and sterile. The specific mode of action of *Btk* on caterpillars is depicted in Figure 2.2. The *Btk* toxins mainly affect the anterior zone of the midgut of lepidopteran larvae. The midgut of the lepidopteran larvae is a simple, tubular epithelium that dominates the internal architecture of the insect. The tissue is composed of two major cell types; columnar cells with a microvilli apical border and a unique goblet cell, containing a large vacuolar cavity, linked to the apical surface by an elaborate and tortuous valve. The K^+ pump is located in the apical membrane of the goblet cell, pumping K^+ from the cytoplasm into the cavity and thence to the gut lumen via the valve. This electrogenic K^+ transport is the predominant feature of the larval lepidopteran gut. Lepidopteran larva is characterized by a high blood ratio of K^+ : Na^+. Another important feature of the midgut is that the pH of the lumenal fluid is about 12 (Knowles, 1994 and Kumar et al., 1996).

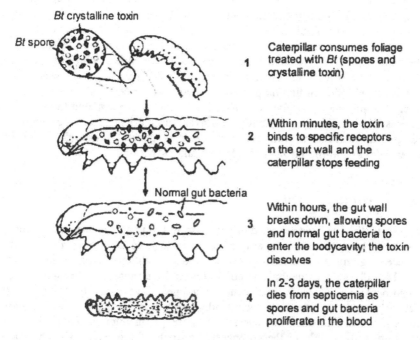

Bt crystalline toxin

Bt spore

Normal gut bacteria

1   Caterpillar consumes foliage treated with *Bt* (spores and crystalline toxin)

2   Within minutes, the toxin binds to specific receptors in the gut wall and the caterpillar stops feeding

3   Within hours, the gut wall breaks down, allowing spores and normal gut bacteria to enter the bodycavity; the toxin dissolves

4   In 2-3 days, the caterpillar dies from septicemia as spores and gut bacteria proliferate in the blood

Figure 2.2   Mode of action of *Bacillus thuringiensis* subsp. *kurstaki* on caterpillars (courtesy: Valent BioSciences Corporation).

2.5   Persistence

On agricultural crops, most of the activity of *Btk* towards target Lepidoptera disappears within 3-4 days after application (Beegle et al., 1981). The low field persistence of this insect pathogen is a major problem regulating its effective use for pest control.

2.5.1   Persistence in water

Field studies have shown that spores of *Btk* persist for some time in fresh water. In laboratory, survival studies for *Btk* in four different types of water i.e. distilled water, tap water, lake water and sea water at 20° C have shown that *Btk* survived for an extended period in all the four tested media. The survival pattern of *Btk* in distilled and tap waters show that approximately 50% of the original cell population died off rather rapidly during the first 20 days period, then declined more steadily and remained relatively unchanged throughout the remaining part of the experimental period from 40 to 70 days.

There is a significant difference in the survival of the *Btk* in lake and sea water. *Btk* was relatively more persistent in fresh water than in sea water. The viability of *Btk* in lake water remained quite stable after 50 days. In contrast, there was a continuous decrease of *Btk* population in sea water. Approximately 90% of the *Btk* population died off after 30 days exposure to sea water. The prolonged survival of *Btk* in lake water was postulated to be due to the presence of higher concentrations of available nutrient which may enhance the growth of bacteria whereas, sea water is generally considered to be bactericidal to non-marine bacteria (Menon and De Mestral, 1985).

2.5.2   *Persistence in soil*

Spores of *Btk* have been reported to persist in soil upto a year or more (Delucca et al., 1981). *Bt* spore can remain viable for long periods of time in soil in the absence of germination-inducing stimuli. Under conditions favoring the growth of soil bacilli, such as neutral pH conditions and the presence of proteinaceous amendments, *Btk* can germinate, compete vegetatively with soil microorganisms, and sporulate successfully to attain levels higher than 1 million spores/gm soil. *Btk* does not survive if spore germination is induced under conditions unfavorable to the competitive growth of soil bacilli (Saleh et al., 1970).

2.5.3   *Persistence on foliage*

Persistence of *Btk* in forest system appears to be more variable. Reported half-life varies from 1 day on oak (*Quercus spp.*) and redbud (*Cercis canadensis*) towards gypsy moth, to 3.5 day in a mixed coniferous forest with activity toward western spruce budworm (*Choristoneura occidentalis*) on some branches persisting 20 days after spraying (Johnson et al., 1995).

2.5.4   *Factors affecting persistence*

The impact of environmental factors (primarily sunlight, temperature, humidity-water) on the field persistence of *Bt* has been discussed in Chapter 1, Section 7.1. However, the sunlight which is probably the most destructive environmental factor affecting the *Btk* is discussed here in detail.

Salama et al. (1983) found that one day of the direct sunlight could inactivate over 90% of *Btk* spores on potted white spruce. The trees themselves in the dark can inactivate 78% of the spores in 14 days. They also reported that *Btk* spores had half-lives between 75 and 256 hr on cotton leaves, not due to high temperatures, but rather due to the effect of ultraviolet radiation. Spore viability of *Btk* was reduced 50% after 30 minutes exposure to simulated sunlight. Endotoxin activity also was reduced, however, it required about 8 times more light exposure (3.8 h) to obtain a 50% loss in insecticidal activity (Ignoffo, 1992). Wavelengths in the 300-380 nm range of the solar spectrum are largely responsible for loss of toxicity in purified *Btk* HD-1 and HD-73 crystals. Sunlight radiation has been shown to cause tryptophan destruction in protein crystals of *Btk* HD-1 and NRD-12 (Pozsgay et al., 1987).

2.6   Safety and Ecotoxicological Effects

The safety of *Btk* is not only beneficial environmentally but also leads to other practical advantages. Unlike most other insecticides, *Btk* does not require special protective clothing, there is no waiting period before re-entering the field, and it may be applied up to the day of harvest. Furthermore, it can be used for aerial spraying of residential areas for control of gypsy moth, without fear of harm to human or pets (Spear, 1987).

2.6.1   Mammalian safety of Btk

Btk has not demonstrated evidence of toxicity, infectivity, irritation or hypersensitivity to mammals. Research workers, manufacturing staff or users have observed no allergic responses or health problems. Human volunteer ingestion and inhalation of *Btk* led to no ill effects. No toxicity in mice, rats or dogs has been demonstrated with single dosage of *Btk* technical up to 10,000 mg/kg of body wt. Thirteen week dietary administration of technical material to rats at dosages of 8,400 mg/kg/day produced no toxic effects. Two years chronic dietary administration of technical material to rats at 8,400 mg/kg/day produced no tumorigenic or oncogenic effects.

No corneal opacity was observed in rabbits treated with 0.1 ml of *Btk* technical. There is also no evidence of sensitization in guinea pigs treated with repeated subcutaneous injections of *Btk* technical material.

2.6.2   Toxicity / pathogenicity to bird and fish

Studies show that *Btk* is not toxic or pathogenic to fish or avian species. *Btk*, when fed to rainbow trout (*Oncorhynchus mykiss*), blue gills, bobwhite quail and for 14 days to chickens, did not produce toxicity to these species.

An avian oral pathogenecity and toxicity study in the mallard duck (LD_{50} >200 gms *Btk*/kg) indicates low toxicity of *Btk* to mallard. Similarly, another avian oral pathogenecity and toxicity study in the bobwhite quail, (LD_{50} >10 gms *Btk*/kg) has shown no adverse effects and autopsy of the bird revealed no pathology attributable to *Btk*.

In the laboratory, when *Btk* was added to water containing the marine fish (*Anguilla anguilla*), no observable toxic or pathogenic effects was observed (Burges, 1981 and Product brochure of DiPel® by Abbott, 1991).

2.6.3   Effects on non-target organisms (NTOs)

Btk demonstrated little or no observable toxicity to non-target organisms in both controlled testing and actual field usage. No significant effect of *Btk* on Zooplankton including rotifers, coperods, cladocerans, phantom midges (*Chaoborus sp.*) and particularly on *Daphnia sp.* was found in 3 months study. *Btk* has no toxic effects on microcrustaceans (Coperoda, Ostracado), mites (Hydracarnia) and insects (Diptera, Heteroptera, Ephemeroptera, Odonata and Coleoptera).

When exposed to concentration of *Btk* equivalent to the worst-case field situation, none of larvae of Trichoptera, Plecoptera, Ephemeroptera, Megaloptera and Diptera species were found to be susceptible, except *Simulium vittatum*. No toxic effect was observed on mussels (*Mytilus edulis*); oysters (*Crassostrea gigas* and *C. virginica*); common periwinkle (*Littorina littorea*); freshwater shrimps (*Crangon crangon*) and the brine shrimp (*Artemia salina*), when exposed to *Btk* in aquaria at concentration of 10 to 400 mg/l for 96 h, brine shrimp was the only species found to be susceptible to the *Btk* in this study (Melin and Cozzi, 1990 and Johnson et al., 1995).

2.6.3a   Effect on parasites

Parasites being important regulators of insect pest populations have been extensively tested for sensitivity or susceptibility to *Btk*. The following observations have been made:

(i)   *Btk* did not affect the parasitism of tachinid species such as *Blepharipa scutellata* and *Parasitigena agilis* and there is a reported case of increase in parasitism by the tachinids, *Compsilura concinnata* and *Blepharipa pratensis*.

(ii)   When washed spores and crystals of *Btk* ($5x10^7$ spores + crystals per ml) were fed to adult *Trichogramma cacoeciae*, no mortality or reduced capacity to parasitize was observed after 7 days feeding (Hassan et al., 1983).

(iii)   No decrease in the percentage of parasitism of aphids by *Diaretiella rapae* was found on collards treated with *Btk*.

(iv)   On aerial spray of *Btk*, there was increase in the percentage of parasitism of gypsy moth larvae by *Cotesia melanoscelus* and *Phobocampe unicincta*. (Ticehurst et al., 1982; Weseloh et al., 1983, and Webb et al., 1989).

2.6.3b   Effect on predators

It is a matter of concern if a beneficial insect predator may become intoxicated or infected when feeding upon a pest species that has ingested *Btk* spores and/or crystals. The following observations have been made:

(i)   When lethal quantities of *Btk* was fed to larval cabbage looper *(Tricoplusia ni)* and just prior to death, these larvae were offered to young chinese praying mantids (*Tenodera aridifolia* ssp. *sinensis)*, it was observed that mantids were not susceptible to spore/crystal mixtures in an intact insect host.

(ii)   When striped earwig (*Labidura riparia*) an important insect predator of lepidopteran larvae is treated with *Btk* at rates equivalent to 10 times the normal field application rate, no mortality was observed.

(iii)   *Btk* has shown an effective control of lepidopteran pest species with no detrimental effect on nymphs or adults of spined stiltbugs *(Jalysus spinosus)* important predators on lepidopteran eggs, particularly those of tobacco budworm (*Heliothis virescens)*, during a 2 months long study.

(iv)   No toxic effect has been observed on the spined soldier bug (*Podisus maculiventris*) following forest spray of *Btk* on the oak leaf caterpillar (*Heterocampa manteo*) (Melin and Cozzi, 1990).

2.6.3c   Effect on beneficials

No adverse effect of *Btk* on beneficial arthropods, predators or parasites has been observed during laboratory and field studies. Those studied include predaceous bugs, big eyed bugs, damsel bugs, assain bugs, lacewings, lady bird beetles, soft winged flower beetles, parasitic wasps, paper-nest wasps, etc.

i)   Effect on honey bees: *Btk* has no toxic effects on honey bees and when fully sporulated culture of *Btk* was fed to adult bees at conc. of $1x10^8$ spores + crystals per bee over a 7 d period, no harmful effect was noted. Honeybees foraging treated areas are not harmed by *Btk* use (Bailey,1971 and Melin and Cozzi,1990).

ii) Effect on silkworms: *Btk* strains consisting of Cry1Aa toxin are recognized to be toxic to silkworm. A *Btk* strain in which spores are inactivated, Toarow CT® (Toagosei Co. Ltd.), is of low toxicity to the silkworm (Navon, 1993).

iii) Effect on earthworms: Earthworms are of great importance in most ecosystems especially in forests. It is therefore highly desirable that the pesticides used should not endanger the earthworms. Various field and laboratory studies indicate no toxic effects of *Btk* on earthworm. When *Btk* at conc. of 30 g/M^2 is applied to small field plots, no adverse effects on the earthworms population was seen and no dead or diseased worms were found in the treatment areas even after 2 months.

In another field experiment with a normal concentration of *Btk* formulation (3600 g/ha or about 1 billion spores/M^2) and 100 times higher concentration (about 1 trillion spores/M^2), no adverse effect of *Btk* was found against earthworms within 7 weeks. During the experiment no evident difference in the density of snails, forficula, myriapodes and woodlice was observed (Benz and Altwegg, 1975 and Melin and Cozzi, 1990). As *Btk* is non-toxic to bees, and because it does not harm predatory insects, it is ideally suited to integrated pest management programs.

3   *Bacillus thuringiensis* subsp. *israelensis*

Bti was isolated in 1976 by Goldberg and Margalit following screening of various isolates from soil samples taken from known mosquito breeding sites in Negev desert of Israel (Goldberg and Margalit, 1977). In 1978, H. de Barjac determined the serotype designation for *Bti* as H-14 at Louis Pasteur Institute at Paris using the standard flagellar technique. *Bti* was found to be fairly toxic to dipteran larvae (especially that of Nematocera), including mosquitoes, blackflies, hornflies and stable flies in the larval stage, but did not have any adverse effect on animals, plants or other insects and was found to be non-toxic to lepidopteran larvae. Mosquitoes and blackflies are not only bothersome but also represent a serious risk to public health for they are vectors of a multitude of diseases of man and animals through transmission of pathogenic viruses, bacteria, protozoa and nematodes.

Thus, *Bti* has turned out to be of considerable importance especially in the tropics, because of its potential as biological insecticide against *Anopheles*, *Aedes* and *Culex* species and *Simulium damnosum*, the vectors of devastating diseases such as malaria, yellow fever, filariasis and river blindness, respectively.

3.1 Characteristics

Bti serotype H-14 is an aerobic, gram-positive, rod shaped, spore forming bacteria. During sporulation, it synthesizes a cytoplasmic parasporal inclusion body (or crystal). The *Bti* crystal is composed of at least four major protein toxins, (a) Cry4A, 134 kDa (b) Cry4B, 128 kDa (c) Cry11A (formerly designated as CryIVD), 72 kDa and (d) Cyt1A, 27 kDa and another minor component Cry10A (formerly designated as CryIVC), 78 kDa. The 27 kDa protein is responsible for cytolytic activity and the higher molecular mass proteins are responsible for dipteran activity and nematocerous toxicity (Bozsik et al., 1993).

3.2 Bti *Crystal Proteins and Genes*

3.2.1 Parasporal inclusion of Bti

Most of the mosquito-active strains of *Bt* produce spherical or irregular shaped parasporal inclusions. The parasporal body of *Bti* is basically spherical and averages about 1 μm in diameter, ranging from 0.7 to 1.2 μm (Figure 2.3).

The toxicity of *Bti*'s parasporal body varies considerably depending on whether it is intact or solubilized and on how it is assayed. When ingested, either intact or solubilized, the parasporal body is toxic to mosquitoes, black flies, and several other nematocerous dipterans. In addition to being toxic to mosquitoes, the solubilized parasporal body is cytolytic to a wide range of vertebrate and invertebrate cells, including erythrocytes, and toxic to mice if injected. Thus, *Bti* is characterized as being mosquitocidal, cytolytic, hemolytic, and even neurotoxic. The cytolytic activity of the solubilized parasporal body is attributed to a 25-kDa protein that is cleaved from the 27-kDa protein when the parasporal body is solubilized under alkaline conditions (Thomas and Ellar, 1983 and Armstrong et al., 1985).

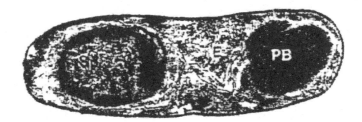

Figure 2.3   Electron micrograph of the parasporal body of *Bti* developing spore (Sp)
and parasporal body (PB) during sporulation; E, exosporium (Source:
Federici et al., 1990).

Several subspecies of *Bt* other than *israelensis* have been reported to
produce parasporal bodies containing proteins toxic to mosquitoes. These
subspecies include *morrisoni, medellin, jegathesan, kyushuensis,
darmstadiensis, shandongiensis, canadensis* and *galleriae*. The parasporal body
produced by the isolate of *Bt* subsp. *morrisoni* PG-14, contains same proteins as
Bti as well as an additional protein of 144 kDa (Federici et al., 1990 and Koni
and Ellar, 1994).

3.2.2   Bti *encoding genes*

The *Bti* contains a 72-mDa plasmid from which several crystal protein genes
(*cry4A, cry4B, cry10A and cry11A*) and the *cyt1A* gene have all been isolated.
The architecture of *cry4A* and *cry4B* is similar to that of the *cry1* genes. They
also encode ca. 130-kDa proteins, which are proteolytically converted into
smaller toxic fragments of 53-67 kDa (Chunjatupornchai et al., 1988). The
cry10A gene encodes a protein with a predicted molecular mass of 78 kDa. The
cry11A gene encodes a 72-kDa protein, which is a major component of the *Bti*
crystals (Federici et al., 1990). This crystal protein, unlike all other known *cry*-
encoded proteins, is proteolytically converted into an active fragment of ca. 30
kDa. The 27-kDa protein encoded by *cyt1A* shows no sequence homology to the
other crystal protein genes (Hofte and Whiteley, 1989).

3.3   Insecticidal Activity

The *Bti* is insecticidal against at least 72 species of mosquitoes (Culicidae), at least seven species of blackflies (Simulidae) and other dipterans, including chironomid larvae, sciarid flies and tipulids. *Bti* toxins have also been shown to possess larvicidal activity against nematodes (Bone et al., 1988). Upon solubilization, the *Bti* toxin lyses erythrocytes and cultured mammalian cells as well as cultured insect cells, whereas solubilized Lepidoptera-active toxins display no corresponding cytotoxicity. Various species of mosquitoes have been noted to exhibit different degrees of susceptibility to a given preparation or formulation of *Bti*. In general, *Culex* mosquito larvae are more susceptible than *Aedes* larvae, followed by *Anopheles* larvae (Mulla, 1986).

In *Bti*, three major anti-dipteran toxins ranging from 68 to 135 kDa (Cry4A, Cry4B, and Cry11A) contribute to the overall toxicity of intact crystals in a synergistic manner (Park et al., 1995). Cyt toxins are hemolytic and cytolytic *in vitro* and are specifically active against dipteran larvae *in vivo* (Bozsik et al., 1993 and Guerchicoff et al., 1997). The Cyt1A toxin displays greater capacity to synergize the action of the Cry4 and Cry11 toxins of *Bti* (Hurley et al., 1987; Gill et al., 1992 and Poncet et al., 1995).

Some of the commercial products are based on two different strains of *Bti*, i.e. VectoBac (Abbott) and Aquabac (Becker Microbial) based on strain HD-507 and Teknar (Thermo Trilogy) based on strain HD-567.

3.3.1   Factors affecting Bti activity

The efficacy of *Bti* in the field is influenced by many factors, such as species of mosquitoes, vegetation cover, extent of water pollution, water flow rate, depth of water, and prevalence of natural enemies.

The commercial formulations of *Bti* available now, can yield 95-100% control of *Culex*, *Aedes* and *Psorophora* larvae at the field rates of 0.11 to 0.5 kg/ha of the primary products. Double and triple of these rates are necessary for controlling *Anopheles* larvae. Rates of application have to be doubled or tripled in treating deep, polluted or vegetated breeding sources (Mulla, 1986).

Various environmental factors such as temperature and pH of water influence the efficacy of biolarvicide. *Bti* displays more activity at higher temperature against *Anopheles* spp. and *Culex* spp. The activity enhancement for *Anopheles stephensi* and *Anopheles culicifacies* from 21°C to 31°C was found to be of the order of 3- and 5-fold respectively (Mittal et al., 1993a) (Table 2.3). On the other hand, at high water pH effectiveness of *Bti* is greatly reduced (Table 2.4).

Table 2.3   Effect of temperature on the activity of *Bti* against *Anopheles* spp. (Adapted from Mittal et al., 1993).

Temperature °C	*LC$_{50}$ Values (mg/l) after 40 hr exposure*			
	Anopheles stephensi		*Anopheles culicifacies*	
	LC$_{50}$	^{31}LC$_{50}$/^{21}LC$_{50}$	LC$_{50}$	^{31}LC$_{50}$/^{21}LC$_{50}$
21°	0.25		0.8	
31°	0.076		0.17	
		3.3		4.7

3.4   Mode of Action

The mode of action of *Bt* is given in detail in Section 6.0 of Chapter 1. The Cyt toxin is a unique component of *Bti* that does not share homology with the Cry toxins. The binding receptors are different for Cry and Cyt toxins. While Cry toxin binds to midgut glycoprotein receptor, the Cyt toxin bind to unsaturated phospholipid. Although both toxins share a common cytolytic mechanism involving colloid-osmotic lysis, the actual mechanism of pore formation may be different. The gut of mosquito larvae is a simple cuboidal epithelium, with no goblet cells. The *Bti* toxin mainly affects the posterior zone of the midgut of dipteran larvae, as against *Btk* toxin that affects anterior zone of the midgut of lepidopteran larvae (Lakhim-Tsror et al., 1983; Knowles and Ellar, 1987; Gill et al., 1992; and Knowles, 1994).

Table 2.4   Effect of water pH on the larvicidal activity of *Bti* (Source: Malaria Research Center, New Delhi, 1993, Progress Report of biolarvicides in vector control)

Mosquito species	Concentration mg/l	*Percent mortality of mosquitoes*			
		pH			
		3.5	7.5	9.5	10.5
Anopheles stephensi	0.5	78	90	94	8
Culex quinquefasciatus	0.25	100	98	88	12

3.5   Persistence

Bti is compatible with the environment and is particularly suited for application in areas that are experiencing resistance development among the target species to organophosphate larvicides. It does not persist in soil or in aquatic environments and its residual activity and persistence in the environment are limited to only a few days. Over time, the *Bti* crystals gradually settle to the bottom of aquatic environments where they become inactivated through binding with soil particles or are used as a protein source by other non-target organisms.

3.6   Safety and Ecotoxicological Effects

After more than a decade of use, *Bti* remains a safe alternative to chemical pesticides used in the aquatic environment. No adverse incidents have ever been reported following its use and there have never been any toxic incidents involving fish, wildlife or man.

3.6.1   Mammalian safety

The toxicological studies of *Bti* were performed by experimental exposure of various animals (mice, rats, guinea pigs, and rabbits) and indicated the absence of acute and prolonged toxicity. All acute toxicity tests gave negative results through a variety of pathways - subcutaneous and intra-peritoneal injection, gavage, inhalation, percutaneous applications, scarification of the skin, and ocular inoculations - using mean doses of about 10^7 to 10^8 bacteria per animal. There was also no anaphylactic shock obtained in guinea pigs, and successive passages in mice of *Bti* did not induce any virulence. In investigating sub-acute oral toxicity in mice and rats, through repeated administration of *Bti* for a period of three weeks, during which the equivalent of about 10^{11} to 10^{12} bacteria per animal was ingested, results were negative. These studies in mice, rats, guinea pigs, and rabbits confirmed that mammals are highly tolerant of the organism. Animals exposed to *Bti* rapidly eliminated the organism from their systems and tissues showed no evidence of multiplication. In summary, neither pathological symptoms, nor diseases, nor mortality have been observed (Siegel and Shadduck, 1990a). *Bti* has also been shown to be safe to amphibians and data indicate that these organisms would not be affected at operational dosages (Paulov, 1985).

All the published data indicate that *Bti* is not a significant toxicant using conventional routes of exposure. Thus, *Bti* is avirulent and non-pathogenic and can be safely used in environments in which human exposure is likely to occur including exposure to immuno-deficient humans (Siegel, Shadduck and Szabo, 1987 and Siegel and Shadduck, 1990b).

3.6.2   Effect on non-target organisms

Bti is considered environmentally benign because tests on most non-target aquatic organisms have not resulted in discovery of deleterious effects. However, several families of Nematocera besides Culicidae and Simuliidae, such as the non-target Blephariceridae and some Chironomidae, are susceptible to *Bti* (Painter et al., 1996).

Selectivity in target and non-target nematoceran Diptera is the result of a combination of filter-feeding behavior, alkaline gut pH, and presence of certain proteolytic enzymes in the gut that activate the toxin. These factors enable the capture, dissolution, and activation of the toxin containing parasporal inclusion bodies (Lacey and Mulla, 1990). The absence of any of these factors greatly reduces or nullifies the susceptibility of a given species to the toxic effects of *Bti*. For example, filter feeding Ephemeroptera, Trichoptera, and others may actually ingest the toxin, but without subsequent activation by pH and proteolytic enzymes, the crystal remains harmless to these organisms. Some Nematocera may be susceptible but their method of feeding (eg. predator, scraper, gatherer, etc.) reduces the amount ingested and hence the exposure is probably below toxic limits (Cibulsky and Fusco, 1987).

It has been found that *Bti* has no noticeable adverse effects on Odonate naiads, dragonfly and damselfly naiads (Ali and Mulla, 1987). *Bti* minimally affects predacious mosquito species of the genus toxorhynchites. There were no adverse effects on crustacea by the operational dosages of *Bti*.

Thus, from the data obtained in laboratory and field tests, it can be said that the use of *Bti* has minimal if any impact toward non-target organisms in the aquatic environment.

4   *Bacillus sphaericus*

Bacillus sphaericus, a cogeneric species of *Bti*, also produces parasporal proteinaceous toxin crystals during sporulation, which are generally active against *Culex* and *Anopheles* species. Kellen and Meyers first isolated in 1964, a *Bacillus sphaericus* strain active against mosquito larvae from moribund fourth instar larvae of *Culiseta incidens*. Kellen's strain was later designated as strain K. This strain had a low order of larvicidal activity because of which it could not be used in mosquito control. The identification of the strain SSII-1 in India by Samuel Singer in 1973, renewed interest in *B. sphaericus*. But only after the isolation of strain 1593 in Indonesia from dead mosquito larvae, which exhibited a much higher level of mosquitocidal activity, was the potential of *B. sphaericus* as a biological control agent taken seriously. In 1980, Wickremesinghe and Mendis isolated what they called strain MR-4. This strain has since been given the designation 2297. This strain along with strains 1593 and 2362, remain some of the principal candidates of field interest among a growing list of 30-50 strains of *B. sphaericus* that are available for use (Singer, 1990 and Charles et al., 1996). *B. sphaericus* has been found to be innocuous to mammals. It also shows a recycling potential under field conditions. In addition, with the discovery of increasingly toxic strains, this biological insecticide seems very promising for control of mosquitoes.

4.1   *Characteristics*

Bacillus sphaericus is a gram-positive, mesophilic, strictly aerobic, round spore former saprophytic bacterium (Figure 2.4). The spore size is around 0.5-1.0 μm. *B. sphaericus* is found commonly throughout the world in soil and aquatic environments.

B. *sphaericus* toxin remains extremely stable under optimal storage conditions of neutral pH and 4°C and fairly stable at room temperature, however

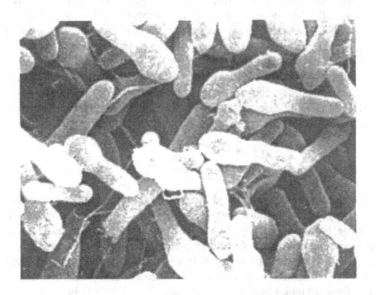

Figure 2.4a   Scanning micrograph of *B. sphaericus* 1593 (Kim and Lee, 1984).

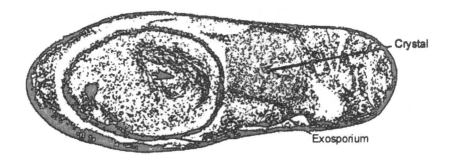

Figure 2.4b   Electron micrograph showing crystal protein in a sporulated cell
of *Bacillus sphaericus* (Charles et al., 1988)

a steady decline in larvicidal activity of a primary powder occurs when stored
out of a desiccator and subjected to temperatures 13-30° C. Toxin that is
exposed to high pH (10.8) will become denatured immediately. A rapid decline
in spore viability, on the other hand, is observed during storage under acidic
conditions. Although viable spores are not required for larvicidal activity, they
are required for persistence of activity due to recycling (Lacey, 1990).

　　　These bacteria are also characterized by negative reaction to most of
the traditional phenotypical tests used for the classification and identification of
bacilli. This largely arises from their obligately aerobic physiology and inability
to use sugars as source of carbon and energy. The findings of various genetic
and biochemical studies indicate that phenotypically *B. sphaericus* is
heterogeneous. Both insecticidal (1593, 2362, 2297) and non-insecticidal (SSII-
I) strains of *B. sphaericus* are known (Russell et al., 1989 and de Barjac, 1990).

4.2   Toxins Produced by B. sphaericus

The entomocidal *B. sphaericus* strains synthesize two types of toxins, binary
toxin (Btx) and mosquitocidal toxin (Mtx). Many high-toxicity strains
synthesize both Btx and the Mtx, while others synthesize only the binary
toxin. Low-toxicity strains synthesize either none of the two toxins or only the
Mtx. Both the toxins differ in their compositions and time of synthesis.
Thereis also evidence for presence of additional toxin genes encoding novel
type of mosquitocidal toxins of 31.8 kDa and 35.8 kDa (Liu et al., 1996 and
Thanabalu and Porter, 1996).

4.2.1 Binary toxin (Btx)

The binary toxin is present in all highly active strains and accumulates during the early stages of sporulation and forms a small crystal in the mother cell. The crystal comprises of equimolar amounts of two proteins of 41.9 and 51.4 kDa, often referred to as the 42 and 51-kDa toxin respectively (Baumann et al., 1991). Both the proteins are needed for larval toxicity. Strains which synthesize the binary toxin are referred to as high toxicity strain because of the acute toxicity conferred by the large amounts of protein in the crystal (Priest et al., 1997).

The amino-acid sequences of 42 kDa and 51 kDa protein are not similar to those of any other bacterial toxins, including those produced by Bti. Therefore, they constitute a separate family of insecticidal toxins. Subcloning experiments have shown that 42 kDa protein alone is toxic for mosquito larvae Culex pipiens, although the activity is weaker than that of crystal containing both proteins. In contrast 51 kDa protein alone is not toxic, but its presence enhances the larvicidal activity of 42 kDa protein, suggesting synergy between the polypeptides. The larger protein is postulated to act as a binding protein enabling the entry of the 42 kDa protein into the midgut cells of the larval gut. (Nicholas et al., 1993 and Charles et al., 1996).

4.2.2 Mosquitocidal toxin (Mtx)

Mosquitocidal toxin (Mtx) is unrelated to Btx. It is synthesized during vegetative growth phase and is proteolytically degraded as the cells enter the stationary phase. Three types of Mtx toxins have been described, Mtx1, Mtx2 and Mtx3 with molecular mass of 100, 30.8 and 35 kDa respectively. These toxins do not display any similarity to each other or to the crystal proteins or any other insecticidal proteins (Baumann et al., 1991).

4.3 Classification

With the increasing number of isolated strains of B. sphaericus, it became apparent that there is a need to differentiate among the strains, as difficulty arises in their identification and differentiation - first among the atoxic and toxic strain and second from each other within the toxic group. There are few phenotypic differences between atoxic and toxic strain other than insecticidal activity. Moreover, the phenotypic characterization that is used in Bacillus taxonomy and which has proven useful for differentiation of B. thuringiensis strains, fail to distinguish between B. sphaericus isolates.

Numerous methods have been tried to classify this very heterogenic species. Classification via bacteriophage typing (Yousten, 1984) and flagellar

(H) serotyping (de Barjac, 1990), similar to that for *B. thuringiensis* strains, have been used. In early studies, toxicity and serotypes or phage-types appeared to correlate, high toxicity strains were classified largely in serotype 5a5b, phage-group 3; serotype 6, phage-group 3 and serotype 25, phage group 4. Low toxicity strains were classified in serotypes 1a, phage-group 1. Strains showing an average toxicity were classified in serotype H26a26b, and serotype H2a2b, phage-group 2. The exception to correlation is the phage-group 4 with less-toxic strains 2173 and 2377 and highly toxic 2297. The discovery of high-toxicity strains from Ghana belonging to serotypes 3, 6 and 48 confused the supposed correlation between serotypes and suggested that the relationships may not be so straightforward (Table 2.5) (de Barjac, 1990 and Priest et al., 1997). Some of the recently developed techniques including numerical classification, ribotyping, cellular fatty acid composition analysis and random amplified polymorphic DNA analysis have been used for the classification which also indicate that most of the pathogenic groups are recovered in a few groups (Charles et al., 1996).

Binary toxin genes were detected in chromosonal DNA of strains from serotypes 1a, 3, 5a5b, 6, 25 and 48 but not in strains from serotypes 2a2b, 9a9c or 26a26b (Table 2.5). Mosquitocidal toxin genes were detected in strains from serotypes 1a, 2a2b, 5a5b, 6, 9a9b, 25 and 48 and is absent from strains of several serotypes, notably all those of serotype 3 and most of serotype 2a2b. Overall there was little correlation between the serotype and toxin gene distribution except for serotype 5a5b strains. Strains with serotypes 1a, 2a2b, 3, 6, 25 and 48, all showed heterogeneity of the toxicity genotype to various degrees. Only strains of serotypes 5a5b (highly toxic) and 26a26b (weakly toxic) were consistent in their toxicity pattern. Thus, although serotyping provides a useful framework for strain identification, its reliability as a predictive tool must be treated cautiously (Priest et al., 1997).

4.4   Insecticidal Activity

B. sphaericus formulations have been reported to be highly insecticidal against larvae of *Culex* and certain species of *Anopheles* but not against *Aedes aegypti*. The toxicity of *B. sphaericus*, particularly against larvae of *Anopheles* species, isgenerally low as compared to that of *Bti*, and the mortality is obtained in 48 to 72 hrs. But it has been reported to persist and recycle in the treated habitats. However, *B. sphaericus* has greater biological activity than *Bti* against *Culex* larvae, producing 95-100% control of *Culex* larvae at the rates of 0.1 to 0.2 kg/ha in clear water. This rate of application has to be increased in polluted, vegetated or deeper bodies of water (Davidson et al., 1981 and Mulla, 1986).

Table 2.5   Classification pattern and distribution of Btx and Mtx genes in *B. sphaericus* (Ref.: de Barjac, 1990; Charles et al., 1996; Zahner and Priest, 1997 and Priest et al., 1997)

Strain	Origin	Btx gene	Mtx gene	Phage type	Sero type
K	United States	No	Yes	1	1a
SSII-1	India	No	Yes	2	2a2b
1889	Israel	No	Yes	2	2a2b
1883	Israel	No	Yes		2a2b
IAB 881	Ghana	Yes	No	NR	3
LPI-G	Singapore	Yes	No	8	3
1593	India	Yes	Yes	3	5a5b
1691	El Salvador	Yes	Yes	3	5a5b
2013.6	Romania	Yes	Yes	3	5a5b
2362	Nigeria	Yes	Yes	3	5a5b
2500	Thailand	Yes	Yes	3	5a5b
BSE 18	Scotland	Yes	Yes	3	5a5b
1593M	Indonesia			3	5a5b
IAB 59	Ghana	Yes	Yes	3	6
COK 31	Turkey	No	Yes	8	9a9c
2297	Sri Lanka	Yes	Yes	4	25
2173	India	No	No	4	26a26b
2377	Indonesia	No	No	4	26a26b
IAB 872	Ghana	Yes	Yes	3	48

The larvicidal activity of the microbial agent depends not only on the strain's intrinsic properties, but also on the species of mosquito and can thus vary within a genus. Toxicity levels vary among the larvicidal serotypes and even within the same serotype (de Barjac, 1990).

A few commercial products based on *B. sphaericus* strains are listed in Table 2.6.

Table 2. 6   Some commercial formulations of *B. sphaericus*

Strain	Trade Name	Manufacturer
B-101	Spherix	Biological Preparations, Berdsk, Russia
2362	Spherimos	Biochem Products
362	VectoLex	Abbott Labs

4.4.1   High temperatures enhance toxicity

The toxicity of *Bacillus sphaericus* against *Anopheles* larvae was found to get enhanced considerably at higher temperature Table 2.7). A *B. sphaericus* formulation in a laboratory study showed a very large enhancement in toxicity at 31°C as compared to toxicity at 21°C; a nearly 250-fold increase against *Anopheles stephensi* and 100 fold increase against *Anopheles culicifacies* (Mittal et al., 1993a). Earlier, Subramoniam et al. (1981) had reported a similar observation of activity enhancement of *B. sphaericus* against *An. subpictus*. In comparison, a *Bti* formulation showed a relatively modest enhancement in toxicity with temperature (see Table 2.3). The effect of temperature on the activity of *B. sphaericus* against *Culex* species was relatively modest.

It is known that *B.sphaericus* toxin binds to midgut cells of *Anopheles* larvae with low affinity as compared to *Culex* larvae, where it binds with high affinity. Perhaps, at higher temperature, in addition to enhancement of larval

Table 2.7   Effect of temperature on the activity of *Bacillus sphaericus* (Adapted from Mittal et al., 1993a)

Temperature	LC$_{50}$ values (mg/l) after 40 hr exposure			
°C	Anopheles stephensi		Anopheles culicifacies	
	LC$_{50}$	LC$_{50}^{31}$/LC$_{50}^{21}$	LC$_{50}$	LC$_{50}^{31}$/LC$_{50}^{21}$
21°	10.2		>48	
31°	0.04		0.48	
		255		>100

feeding rate, binding characteristics of *B.sphaericus* to larval midgut cells of *Anopheles* species also improve significantly (Figure 2.5).

4.5   Mode of Action

Following ingestion of the spore-crystal complex by a susceptible larva, the protein-crystal matrix quickly dissolves in the lumen of anterior stomach through the combined action of midgut proteolytic enzymes and high pH. It is followed by activation of the 51 and 42 kDa toxin of the binary toxin to 43 and 39 kDa toxins respectively, and binding of the activated toxins to specific receptor molecules. The role of the N-terminal region of the 51 kDa is to direct the binding of the toxins to specific receptor molecules, and its C terminus is necessary to interact with the 42 kDa toxin. *B. sphaericus* releases the toxin in all species, even in non-susceptible species such as *Ae. aegypti*. Midgut alteration starts as soon as 15 min. after ingestion. *B. sphaericus* toxins cause disruption in midgut epithelial cells, specifically in the regions of gastric caecae and posterior midgut. This leads to mortality within 4 - 48 hr after treatment, depending on doses ingested. However, the mode of action of Mtx toxins is still unknown (Charles et al., 1986).

Studies have identified the various factors that are responsible for differences in susceptibility among mosquitoes to *B. sphaericus*. The lack of susceptibility in *Aedes* mosquitoes (especially *Ae. aegypti*) to *B. sphaericus* is not due to rates of toxin ingestion, dissolution of the toxin, or specificity of the gut proteases, but is due to failure of toxins to bind to midgut epithelial cells,

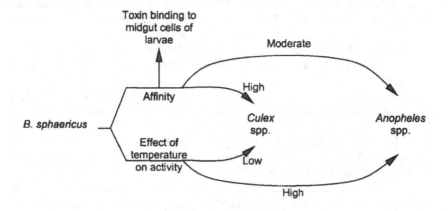

Figure 2.5   Effect of temperature on toxin binding and activity of *B. sphaericus*

since no functional receptors were found in this species. For *Anopheles* species, *B. sphaericus* toxins bind to larval midgut cells but with low specificity and affinity, thus resulting in moderate degree of susceptibility. In *Culex* species the toxins bind with high specificity and high affinity, thus leading to high susceptibility in these species (Rodcharoen and Mulla, 1995).

4.6   Persistence

Field trial experiments have demonstrated that *B. sphaericus* has recycling and persistence ability. This may be due to recycling and multiplication of spores in larval cadavers and certain aquatic situations or may be simply due to long-term persistence of sufficient and accessible toxin or a combination of both (Davidson et al., 1984; Des Rochers and Garcia, 1984). However, the activity and persistence of *B. sphaericus* depends upon the type of strain, its fermentation and formulation methods and also on environmental factors such as water quality and depth, solar radiation, target species, and larval density.

B. sphaericus* can apparently persist in a variety of habitats, including those that are polluted. In shallow containers, in clear water, larvicidal activity has been reported as persisting for nine months. In some extremely enriched and/or deep breeding sites, however, curtailment of larvicidal activity has been reported. Ostensibly, toxin either settled out of the feeding zone of the larvae or ingestion of sufficient quantities was inhibited by excess particulate matter (Lacey, 1990).

In laboratory studies with the SSII-1 isolate, Davidson et al. (1975) reported the germination, growth and sporulation of *B. sphaericus* in cadavers of *Cx. quinquefasciatus* with a subsequent log increase in spores over that ingested by the larvae. Hertlien et al. (1979) retrieved the viable and infective spores of the 1593 isolate several months after introduction into larval habitats. They concluded that *B. sphaericus* is capable of recycling in nature.

Various studies indicate that, under certain conditions, larva to larva recycling occurs and may perpetuate effective control for several months. Amplifications of *B. sphaericus* in larvae may provide significant recycling under field conditions when large numbers of susceptible larvae are present at the time of treatment. Larval density appears to be an important factor in determining the occurrence of recycling of *B. sphaericus* and hence it's effective duration of activity. Larval numbers, in turn may be a function of species and/or environmental conditions (Lacey, 1990).

Although the recycling of *B. sphaericus* may not provide enough spores for long term vector control, the implication that subsequent application dosages in same locality can be lower should be taken into consideration when cost-effectiveness of using *B. sphaericus* for vector control is deliberated (Yap, 1990).

4.7   Safety and Ecotoxicological Effects

B. sphaericus is quite specific, impacting mosquito larvae only, with no adverse effects on mosquito predators. The LC_{50} value of Spherix (a formulated product of *B. sphaericus* strain B-101, serotype H5a5b) against larvivorous fish *Gambusia affinis* and *Poecilia reticulata* and also frog tadpoles and *Mesocyclops* is reported to be greater than 1 g/l. This concentration is approximately 500 and 20,000 times more than that required for effective activity against larvae of *An. stephensi* and *Cx. quinquefasciatus* respectively. Similarly, the LC_{50} values against notonectid bugs, *Enithares indica* and *Anisops sardea* is found to be greater than 100 mg/l and 50 mg/l respectively, which are also approximately 500 and 250 times higher than in *An. stephensi* larvae (Mittal et al., 1993b). *B. sphaericus* is non-toxic to most non-target organisms (Mulla, 1986) and has shown no mammalian toxicity, is safe to fish and wildlife and practically all macro-invertebrate associates of mosquito larvae.

4.8   Comparison of Characteristics of Bti and B. sphaericus

Bti and *Bacillus sphaericus* are two gram-positive soil bacteria that have been successfully used for biological control of blackflies and *Culex* mosquitoes. However, both *Bti* and *B. sphaericus* crystals differ in toxin composition, modes of action, and insecticidal spectra (Table 2.8) (Priest, 1992; Porter et al., 1993 and Porter, 1996).

5   Concluding Remarks

Btk is widely used as a microbial insecticide against lepidopteran pests. Due to its specificity, *Btk* does not disrupt the natural balance between pests and beneficials. This is of particular advantage for forest and other control programs where it is desirable to maintain a natural balance of beneficials to suppress the resurgence of damaging insects. The short half life of *B. thuringiensis* in the field is generally believed to minimize its impact on non-target organisms. On agricultural crops, most of the activity of *Btk* towards target Lepidoptera disappears within 3-4 days after application. In addition to the development of *Btk* isolates as conventional insecticides, transconjugate and genetically engineered strain with novel properties are increasingly coming to the market (see Chapter 3).

Efforts to reduce the populations of dipteran species which transmit tropical diseases such as malaria, filariasis, and onchocerciasis through naturally occurring entomopathogenic bacterial strains, has been receiving increased attention. Over the last two decades, there has been much interest in the toxins

Table 2.8   Comparison of characteristics of *Bti* and *B. sphaericus*

Characteristics	Bt-israelensis	B. sphaericus
Toxin Composition	*Bti* inclusions are made of four major proteins, Cry4A, Cry4B, Cry11A and Cyt1A.	*B. sphaericus* inclusions are made of equimolar amount of proteins of 42- and 51 kDa which act as a binary toxin.
	Protoxin crystals produced during sporulation are deposited along side the spore.	Protoxin surrounded by the exsporium membrane.
Mode of action	*Bti* toxin completely breaks down the larval midgut epithelium.	*B. sphaericus* toxin does not break down completely.
Insecticidal spectra	*Bti* is toxic to *Aedes, Culex, Anopheles, Mansonia* and *Simulium larvae.*	Target spectrum is restricted. It is toxic to *Culex* and *Anopheles* and is poorly or non-toxic to *Aedes* larvae.
	Pathogenic to blackfly larvae.	Not pathogenic to blackfly larvae.
Effect of temperature on activity	Activity enhancement against *Anopheles* and *Culex* larvae at higher temperatures.	Very large activity enhancement against *Anopheline* spp. at higher temperatures.
Persistence and Recycling	*Bti* does not persist for long periods in the environment, especially in polluted water.	More persistent than that of *Bti*, including in polluted water.
	Bti does not recycle.	*B. sphaericus* can recycle under certain environmental conditions.
	Bti crystals are not linked to the spores and they settle more rapidly than *B. sphaericus.*	Toxic agents are closely linked to spore which would then function as a kind of float.
Resistance	Resistance has not been observed even after years of intensive field treatments, probably because of multi-component structure of the crystal.	Potential for development of resistance to binary toxin.

of *Bti* and *B. sphaericus* for mosquito larval control. Their high specificity for their target mosquitoes has an important practical consequence. *Bti* can be regarded, as a natural alternative to chemical insecticides for mosquito control as it is harmless for the non-target organisms of the water fauna. In practice, *Bti* has been used as a replacement for chemical insecticides whenever they have been banned or where insecticide resistance has developed. *Bti* has been widely used in West Africa in a successful vector compaign (the World Health Organization's Onchocerciasis Control Program) against blackflies, *Simulium damnosum* Theobald. There are some reports of *Bti* use in integrated control programs e.g. of mosquitoes, in conjunction with larvivorous fish (see Chapter 10) and other natural predators. The lack of any serious resistance to *Bti* in mosquitoes and blackflies has been interpreted as possibly due to complexity of the multi-toxin (Cry4 and Cry11) inclusions (see Chapter 5). The development of *B.sphaericus* as a microbial control agent of mosquitoes has been somewhat hampered by the success and broader spectrum of *Bti*. Despite that, *B. sphaericus* offers some distinct advantages over its commercially produced counterpart. Most notable is the increased duration of larvicidal activity against certain species of mosquitoes especially in organically enriched (polluted) habitats. However, the future impact of *Bacillus sphaericus* will depend largely on its ability to be more persistent, its recycling ability in field trials, and its possibility for use as a microbial control agent in malaria control programs.

References

1   Ali, C. and M. S. Mulla, 1987. "Effect of two microbial insecticides on aquatic predators of mosquitoes", *Z. Angew. Entomol.*,103, 113.

2   Armstrong, J. L., G. F. Rohrmann and G. S. Beaudreau, 1985. "Delta endotoxin of *Bacillus thuringiensis* subsp. *israelensis*", *J. Bacteriol.*, 161(1), 39-46.

3   Bailey, L., 1971. "The safety of pest-insect pathogens for beneficial insects", in *MicrobialControl of Insect and Mites*, H. D. Burges and N. W. Hussey, eds., Academic Press, pp.491-505.

4   Baumann, P., M. A. Clark, L. Baumann and A. H. Broadwell, 1991. "*Bacillus sphaericus* as a mosquito pathogen - Properties of the organism and its toxins", *Microbiol. Rev.*, 55, 425-436.

5   Beegle, C. C., H. T. Dulmage, D. A. Wolfenbarger and E. Martinez, 1981. "Persistence of *Bt* Berliner insecticidal activity on cotton foliage", *Environ Entomol.*, 10, 400-401.

6   Benz, G. and A. Altwegg, 1975. "Safety of *Bacillus thuringiensis* for earthworms", *J. Invertebr. Pathol.*, 26, 125-126.

7   Bone, L. W., K. Bottijer and S. S. Gill, 1988. "Factors affecting the larvicidal activity of *Bacillus thuringiensis israelensis* toxin for *Trichostrongylus colubriformis* (Nematoda)", *J. Invertebr. Pathol.*, 52, 102-107.

8   Bozsik, A., P. Kiss, F. Fabian, L. Szalay-Marzso and M. Sajgo, 1993. "Insecticidal and haemolytic characterization of the fractions of *Bacillus*

thuringiensis subsp. *israelensis* toxin", *Acta Phytopathol. Entomol. Hungarica*, 28(2-4), 451-460.

9. Burges, H. D., 1981. "Safety, safety testing and quality control of microbial pesticides", in *Microbial control of Pests and Plant Diseases 1970-1980*, H. D. Burges, ed., Academic Press, NY., pp. 738-767.

10. Carlton, B. C. and J. M. Gonzalez, 1985. "The genetics and molecular biology of Bacillus thuringiensis, in *The Molecular Biology of the Bacilli*, Vol. II, D. A. Dubnau, ed., Academic Press, New York, pp. 211-249.

11. Cannon, R. J. C., 1996. "*Bacillus thuringiensis* use in agriculture: A molecular perspective", *Biol. Rev.*, 71, 561-636.

12. Charles, J. F., C. Nielsen-LeRoux and A. Delecluse, 1996. "*Bacillus sphaericus* toxins: molecular biology and mode of action", *Annu. Rev. Entomol.*, 41, 451-472.

13. Chunjatupornchai, W., H. Hofte, J. Seurinck, C. Angsuthanasombat and M. Vaeck, 1988. "Common features of *Bacillus thuringiensis* toxin specific for Diptera and Lepidoptera", *Eur. J. Biochem.*, 173, 9-16.

14 Cibulsky, R. J. and R. A. Fusco, 1987. "Recent Experience with Vectobac for blackfly control: an industrial perspective on future developments in blackflies", in *Ecology, Population Management and Annodated World List*, K. C.Kim and R. W. Merritt, eds., Pennsylvania State University, University Park, pp. 419.

15. Curtis, C. F., 1996. "Delivering biocontrol in the tropics", *Nature Biotechnol.*, 14, 265.

16. Davidson, E. W., S. Singer and J. D. Briggs, 1975. "Pathogenesis of *B. sphaericus* strain SSII-1 infections in *Culex pipiens quinquefasciatus* larvae", *J. Invertebr. Pathol.*, 25, 179-184.

17. Davidson, E. W., A. W. Sweeney and R. Cooper, 1981. "Comparative field trials of *Bacillus sphaericus* strain 1593 and *Bacillus thringiensis* var. *israelensis* commercial powder formulations", *J. Econ. Entomol.*, 74, 350-354.

18. Davidson, E. W., M. Urbina, J. Payne, M. S. Mulla, M. Darwazeh, H. T. Dulmage and J. A. Correa, 1984. "Fate of *Bacillus sphaericus* 1593 and 2362 spores used as larvicides in the aquatic environment", *Appl. Environ. Microbiol.*, 47, 125-129.

19. de Barjac, H., 1990. "Classification of *B. sphaericus* strains and comparative toxicity to mosquito larvae", in *Bacterial Control of Mosquitoes and Blackflies*, H. de Barjac and D. J. Sutherland, eds., London, Unwin Hyman, pp. 228-236.

20. Delucca, A. J., II, J. G. Simonson and A. D. Larson, 1981. "*Bacillus thuringiensis* distribution in soils of the United States", *Can. J. Microbiol.* 27, 865-870.

21. Des Rochers, B. and R. Garcia, 1984. "Evidence for persistence and recycling of *Bacillus sphaericus*", *Mosq. News*, 44, 160-165.

22. Dulmage, H. T., 1970. "Insecticidal activity of HD-1, a new isolate of *Bacillus thuringiensis* var. *alesti*", *J. Invertebr. Pathol.*, 15, 232-239.

23. Federici, B. A., 1993. "Insecticidal bacterial protein identify the midgut epithelium as a source of novel target sites for insect control", *Arch. Insect Biochem. Physiol.*, 22, 357-371.

24. Federici, B. A., P. Luthy and J.E. Ibarra, 1990. "The parasporal body of *Bti*: Structure, protein composition and toxicity", in *Bacterial Control of*

Mosquitoes and Blackflies, H. de Barjac, D. J. Sutherland, eds., Unwin Hyman, London, pp. 16-44.

25. Gill, S. S., E.A. Cowles and P. V. Pietrontonio , 1992. "The mode of action of *Bacillus thuringiensis* endotoxins". *Annu. Rev. Entomol.*, 37, 615-636.

26. Goldberg, L. J. and J. Margalit, 1977. "A bacterial spore demonstrating rapid larvicidal activity against *Anopheles sergentii, Uranotaenia unguiculata, Culex univitattus, Aedes aegypti and Culex pipiens*", *Mosq. News*, 37, 355-358.

27. Guerchicoff, A., R. A. Ugalde and C. P. Rubinstein, 1997. "Identification and characterization of a previously undescribed *cyt* gene in *Bacillus thuringiensis* subsp. *israelensis*". *Appl. Envirn. Microbiol.*, 63(7), 2716-2721.

28. Hassan, S. A., F. Bigler, H. Bogenschutz, J. U. Brown, P. Firth, M. S. Ledieu, E. Naton, PA. Oomen, W. P.J. Overmeer, W. Rieckermann, L. Samsoe-Petersen, G. Viggiani and A. Q. van Zon, 1983. "Results of the second joint pesticide testing programs by IOBC/WPRS-working group", Pesticides and Beneficial Arthropods, *Z. Ang. Entomol.*, 95, 151-158.

29. Hertlein, B. C., R. Levy and T. W. Miller, Jr., 1979. "Recycling potential and selective retrieval of *B. sphaericus* from soil in a mosquito habitat", *J. Invertebr. Pathol.*, 33, 217-221.

30. Hofte, H. and H. R. Whiteley, 1989. "Insecticidal crystal proteins of *Bacillus thuringiensis*", *Microbiol. Rev.*, 53(2), 242-255.

31. Hurley, J. M., L. A. Bulla, Jr. and R. E. Andrews, Jr., 1987. "Purification of the mosquitocidal and cytolytic proteins of *Bacillus thuringiensis* subsp. *israelensis*", *Appl. Environ. Microbiol.*, 53(6), 1316-1321.

32. Ignoffo, C. M., 1992. "Environmental factors affecting persistence of entomopathogens", *Florida Entomol.*, 75(4), 516-525.

33. Johnson, K. S., J. M. Scriber, J. K. Nitao and D. R. Smitley, 1995. "Toxicity of *Bacillus thuringiensis* var. *kurstaki* to three non-target Lepidoptera in field studies". *Environ. Entomol.*, 24(2), 288-297.

34. Keller, B., and G.-A. Langenbruch, 1993. "Control of coleopteran pests by *Bacillus thuringiensis*", in *Bacillus thuringiensis, An Environmental Biopesticide: Theory and Practice*, P.F. Entwistle, J. S. Cory, M. J. Bailey and S. Higgs, eds., John Wiley. Chichester, U.K., pp. 171-191.

35. Kim, Y. H. and H. H. Lee, 1984. "Isolation of sporeless temperature-sensitive mutants of *Bacillus sphaericus*", *HG. J. Gen Eng.*, 1, 15-20.

36. Knowles, B. H., 1994. "Mechanism of action of *Bacillus thuringiensis* insecticidal δ-endotoxin", *Adv. Insect Physiol.*, 24, 275-308.

37. Knowles, B. H. and D. J. Ellar, 1987. "Colloid-osmotic lysis is a general feature of the mechanism of action of *Bacillus thuringiensis* δ-endotoxin with different specificity", *Biochim. Biophys. Acta*, 924, 509-518.

38. Koni, P. A. and D. J. Ellar, 1994. "Biochemical characterization of *Bacillus thuringiensis* cytolytic δ-endotoxins". *Microbiol.*, 140, 1869-1880.

39. Kumar, P. A., R. P. Sharma and V. S. Malik, 1996. "The Insecticidal protein of *Bacillus thuringiensis*", *Adv. Appl. Microbiol.*, 12, 1-43.

40. Lacey, L. A. and M. S. Mulla, 1990. "Safety of *Bacillus thuringiensis* subsp. *israelensis* and *Bacillus sphaericus* to non-target organisms in the aquatic environment", in *Safety of Microbial Insecticides*, M. Laird, L. A. Lacey and E. W. Davidson, eds., CRC Press Inc., Florida, pp. 169-188.

41. Lacey, L. A., 1990. "Persistence and formulation of *B. sphaericus*", in *Bacterial Control* of *Mosquitoes* and *Blackflies*, H. de Barjac and D.J. Sutherland, eds., London, Unwin Hyman, pp. 284-294.

42. Lakhim-Tsror, L., C. Pascar-Gluzman, J. Margalit and Z. Barak, 1983. "Larvicidal activity of *Bti* serovar H-14 in *Aedes aegypti*: Histopathological studies, *J. Invertebr. Pathol.*, 41, 104-116.

43. Liu, J.W., A.G.Porter, B.Y. Wee and T.Thanabalu, 1996. "New gene from nine *B.sphaericus* strains encoding highly conserved 35.8 kDa mosquitocidal toxins", *Appl. Environ. Microbiol.*, 62, 2174-2176.

44. Masson, L., M. Bosse, G. Prefontain, L. Peloquin, P. C. K. Lau and R. Brousseau, 1990. "Characterization of parasporal crystal toxins of *Bt* subsp. *kurstaki* strains HD-1, and NRD-12", in *Analytical chemistry of Bacillus thuringiensis*, L. A. Hickle and W. L. Finch, eds., American Chemical Society, Washington, D.C., pp.61-69.

45. Melin, B. E. and E. M. Cozzi, 1990. "Safety to non-target invertebrates of lepidopteran strains of *Bacillus thuringiensis* and their δ-exdotoxins", in *Safety of Microbial Insecticides*, M. Laird, L. A. Lacey and E. W. Davidson, eds., CRC Press, Florida, pp.149-166.

46. Menon, A. S. and J. De Mestral, 1985. "Survival of *Bacillus thuringiensis* var. *kurstaki* in waters", *Water Air Soil Pollut.*, 25, 265-274.

47. Mittal, P. K., T. Adak and V. P. Sharma, 1993a. "Effect of temperature on toxicity of two bioinsecticides Spherix (*Bacillus sphaericus*) and Bactoculicide (*Bacillus thuringiensis* H-14) against larvae of four vector mosquitoes", *Indian J. Malariology, 30, 37-41.*

48. Mittal, P. K., T. Adak, C. P. Batra and V. P. sharma, 1993b, "Laboratory and field evaluation of spherix a formulation of *B.sphaericus* (B-101) to control breeding of *An. stephensi* and *Cx. quinquifasciatus*", *Indian J. Malariology*, 30, 81-89.

49. Mulla, M. S., 1986. "Role of *Bti* and *Bacillus sphaericus* in mosquito control programs", in *Fundamental and Applied Aspects of Invertebrate Pathology*, R.A. Samson, J. M. Vlak and D. Peters, eds., Foundation of the 4th Int'l Colloq. of Invertbr. Pathol., pp. 494-496.

50. Navon, A., 1993. "Control of lepidopteran pests with *Bacillus thuringiensis*", in *Bacillus thuringiensis, An Environmental Biopesticide: Theory and Practice*, P.F. Entwistle, J. S. Cory, M. J. Bailey and S. Higgs, eds., John Wiley, Chichester, U. K., pp. 125-146.

51. Nicolas, L. C. Nielsen-LeRoux, J.F. Charles and A. Delecluse, 1993, "Respective role of the 42- and 51-kDa component of the *B. sphaericus* toxin overexpressed in *Bacillus thuringiensis*", *FEMS Microbiol. Lett.*,106, 275-80.

52. Painter, M. K., K. J. Tennessen and T. D. Richardson, 1996. "Effects of repeated applications of *Bacillus thuringiensis israelensis* on the mosquito predator *Erythmis Simpliciollis* (Odonata: Libellulidae) from hatching to final instar", *Environ. Entomol.*, 25(1), 184-191.

53. Park, H., H. Dim, D. Lee, Y. Yu, B. Jin and S. Kang, 1995. "Expression and synergistic effect of three types of crystal protein genes in *Bacillus thuringiensis*", *Biochem Biophys. Res. Commun.*, 214, 602-607.

54. Paulov, S., 1985. "Interaction of *Bacillus thuringiensis* var. *israelensis* with developmental stages of amphibians (*Rana temporaria* L.)", *Biologica*, 40, 133.

55.   Poncet, S., A. Delecluse, A. Klier and G. Rapoport, 1995. "Evaluation of synergistic interactions among the CryIVA, CryIVB and CryIVD toxic components of *B. thuringiensis* subsp. *israelensis* crystals", *J. Invertebr. Pathol.*, 66, 131-135.

56.   Porter, A. G., E. W. Davidson and J.W. Liu, 1993. "Mosquitocidal toxins of *Bacilli* and their effective manipulation for biological control of mosquitoes", *Microbiol. Rev.*, 57, 838-861.

57.   Porter, A.G., 1996. "Mosquitocidal toxins, genes and bacteria: the hit squad", *Parasitol. Today*, 12, 175-179.

58.   Pozsgay, M., P. Fast, H. Kaplan and P. R. Carey, 1987. "The effect of sunlight on the protein crystals from *Btk* HD-1 and NRD-12: A Raman spectroscopic study", *J. Invertebr. Pathol.*, 50, 246-253.

59.   Priest, F. G., 1992. "Biological control of mosquitoes and other biting flies by *Bacillus sphaericus* and *Bacillus thuringiensis*", *J. Appl. Bacteriol.*, 72, 357-369.

60.   Priest, F. G., L. Ebdrup, V. Zahner and P. E. Carter, 1997. "Distribution and characterization of mosquitocidal toxin genes in some strains of *Bacillus sphaericus*", *Appl. Environ. Microbiol.*, 63(4), 1195-1198.

61.   Rodcharoen, J. and M. S. Mulla, 1995. "Comparative ingestion rates of *Culex quinquefasciatus* susceptible and resistant to *B. sphaericus*", *J. Invertebr. Pathol.*, 66, 242-248.

62.   Russell, B. L., S. A. Jelley and A. A. Yousten, 1989. "Carbohydrate metabolism in the mosquito-pathogen *Bacillus sphaericus*", *Appl. Environ. Microbiol.*, 55, 294-297.

63.   Salama, H., M. S. Foda, F. N. Zaki and A. Khalafallah, 1983. "Persistence of *Bacillus thuringiensis* Berliner spores in cotton cultivations", *Z. Angew. Entomol.*, 95, 321-326.

64.   Saleh, S. M., R. F. Harris and O. N. Allen, 1970. "Fate of *Bacillus thuringiensis* in soil: Effect of soil pH and organic amendment", *Can.J. Microbiol.*, 16, 677-680.

65.   Siegel, J. P. and J. A. Shadduck, 1990a. "Mammalian safety of *Bacillus thuringiensis israelensis*, in *Bacterial Control of Mosquitoes and Blackflies*", H. de Barjac and D. J. Sutherland, eds., Unwin Hyman, London, pp. 202-217.

66.   Siegel, J. P. and J. A. Shadduck, 1990b. "Safety of microbial insecticides to vertebrates - humans", in *Safety of Microbial Insecticides*, M. Laird, L. A. Lacey and E. W. Davidson, eds., CRC Press Inc., Florida, pp.

67.   Siegel, J. P., J. A. Shadduck, and J. Szabo, 1987. "Safety of the entomopathogen *Bti* for mammals", *J. Econ. Entomol.*, 80, 717-723.

68.   Singer, S., 1990. "Introduction to the study of *B. sphaericus* as a mosquito control agent", in *Bacterial Control of Mosquitoes and Blackflies*, H. de Barjac and D. J. Sutherland, eds., Unwin Hyman, London, pp.221-227.

69.   Spear, B. B., 1987. "Genetic engineering of bacterial Insecticides, in *Biotechnology in Agricultural Chemistry*, H. M. LeBaron, R. C. Honeycutt, J. H. Duesing, J. F. Phillips and M. J. Haas", eds., American Chemical Society, Washington, DC, pp. 204-214.

70.   Subramoniam, A., R. Jamuna and K. Jayaraman (1981). "Temperature dependent activity of mosquito larvicidal factors present in *B. sphaericus* 1593-4 and 1691", *Experientia*, 37, 288-289.

71.   Thanabalu, T. and A.G. Porter, 1996. "A *B. sphaericus* gene encoding a novel type of mosquitocidal toxin of 31.8 kDa, *Gene*, 170, 85-89.

72.   Thomas, W. E. and D. J. Ellar 1983. "*Bacillus thuringiensis* var. *israelensis* crystal of delta endotoxin: Effects on insect and mammalian cells *in vitro* and *in vivo*", *J. Cell Sci.*, 60, 181-19.

73.   Ticehurst, M., R. A. Fusco and E. M. Blumenthal, 1982. "Effects of reduced rates of Dipel 4L, Dylox 1.5 oil and Dimilin W-25 on *Lymantria dispar* (L.) (Lepidoptera: Lumantriidae), parasitism and defoliation", *Environ. Entomol.*, 11, 1058-1062.

74.   Webb, R. E., M. Shapiro, J. D. Podgwaite, R. C. Reardon, K. M. Tatman, L. Venables and D. M. Kolodny-Hirsch, 1989. "Effect of aerial spraying with Dimilin, Dipel, or Gypchek on two natural enemies of the gypsy moth (Lepidoptera: Lymantriidae)", *J. Econ. Entomol.*, 82, 1695-1701.

75.   Weseloh, R. M., T. G. Andreadis, and R. E. B. Moore, 1983. "Field confirmation of a mechanism causing synergism between *Bacillus thuringiensis* and the gypsy moth parasitoid, *Apanteles melanoscelus*", *J. Invert. Pathol.*, 41, 99-103.

76.   Yap, H. H., 1990. "Field trials of *Bacillus sphaericus* for mosquito control", in *Bacterial Control of Mosquitoes and Blackflies*, H. de Barjac and D. J. Sutherland, eds., Unwin Hyman, London, pp. 307-320.

77.   Yousten, A. A., 1984. "Bacteriophages typing of mosquito - pathogenic strains of *Bacillus sphaericus*", *J. Invertebr. Pathol.*, 43, 124-125.

78.   Zahner, V. and F. G. Priest, 1997. "Distribution of restriction endo-nucleases among some entomopathogenic strains of *Bacillus sphaericus*", *Lett. Appl. Microbiol.*, 24, 483-487.

3

Genetically Modified *Bt* Strains and *Bt* Transgenic Plants

1   Introduction

The importance of *Bt* in an environmentally sound insect control program is well accepted. Therefore there is an active interest in search for novel toxins as well as in the discovery of more active strains. The bacterial strains used for all of the early and some of the current *Bt* foliar insecticides are wild type strains, that is, they were found in nature in the form in which they are used to produce the microbial spray. The use of conventional *Bt* insecticides, however has limitation like narrow specificity, low potency and short shelf life. On the other hand, certain combinations of Cry proteins have been shown to exhibit synergistic effects. Therefore, genetic manipulation of *Bt* – to create combinations of genes more useful for a given application than those known to occur in natural isolates - may be desirable. These combinations can be constructed by various approaches that utilize the tools of molecular biology and genetic engineering, as well as conventional microbiological methods.

Several reviews have appeared in the literature covering genetically modified *Bt* strains (Baum et al., 1999; Cannon, 1996; Kumar et al., 1996; Carlton and Gawron-Burke, 1993; and Koziel et al., 1993b) and transgenic *Bt* plants (Huang et al., 1999; Jenkins, 1999; Schuler et al., 1998; Peferoen, 1997 and Estruch et al., 1997). This chapter provides an overview of the developments in both genetically modified Bt strains and Bt crop plants.

2   Novel Bt Strains Through Conventional Genetic Techniques

The manipulation of *Bt* at the molecular level requires techniques for the efficient movement of DNA into and among *Bt* strains. Several methods have been used to move DNA between *Bt* strains and from heterologous species. It is possible to construct strains with increased activity on specific pests by using classical genetic techniques to eliminate unnecessary δ-endotoxins and / or introduce desirable ones.

Multiple δ–endotoxin genes could be introduced into a strain for one or more of the following reasons:

(a)   To increase the toxicity on a particular pest,

(b)   To broaden the host range in cases where one δ-endotoxin can not be identified with acceptable toxicity on multiple insect targets, and

(c)   To manage resistance by including δ-endotoxins that bind to different sites in the insect midgut.

However, the effectiveness of the strain will depend on the biosynthetic capacity of the cell, because introducing more toxins to broaden specificity will usually reduce the amount of each δ-endotoxin present. Regardless of the technique used for rational strain construction, the first and perhaps the most difficult step is to determine the desired insecticidal activity and to identify δ-endotoxins with these properties.

Bt strains present in several currently available bioinsecticides were constructed using the following techniques.

2.1   Conjugation

Bacillus thuringiensis in nature generally contain several different genes that endow different insecticidal activities. *Bt cry* genes usually reside on specific non-essential DNA molecules termed plasmids, although some are chromosomally located also. Plasmids can either be removed (cured) from cells that harbor them or they can be transferred between different strains by a natural mating process called conjugal transfer. It occurs in nature between *Bt* strains and can be replicated in the laboratory producing new *Bt* strains with different spectra of toxicity.

Ecogen Inc., a U.S. Biotechnology company has been actively involved in development of trans-conjugate strains of *Bt* with novel properties that have higher levels of effectiveness and that can act against a broader array of significant insect pests than wild-*Bt* strains. For instance, they combined beetle-active with caterpillar-active *Bt* proteins in the same strain by conjugating *Bt-kurstaki* with *Bt-tenebrionis* to produce a hybrid stain EG 2424 (Foil®). It contains a unique combination of coleopteran- and lepidopteran-active *Bt* ICPs (Cry1Ac (2) and Cry3A) and resultantly has an expanded host range to control Colorado potato beetle in potatoes and European corn borer (Carlton and Gawron-Burke, 1993), However, Ecogen has discontinued Foil® product.

Similarly other commercial products such as Cutlass® and Condor®, which combine the insecticidal properties of both parent strains, have been produced by conjugation process. For their construction, a self-transmissible *cry1A* plasmid was transferred via conjugation from a *Bt-aizawai* strain to a *Bt-kurstaki* recipient strain. On the otherhand, in the products such as Agree® and Design®, a *cry1* plasmid from a *Bt-kurstaki* strain was transferred to a *Bt-aizawai* recipient strain (Baum et al., 1999). The active ingredient of Agree® consists of Cry1Aa, Cry1Ac, Cry1C and Cry1D proteins. Various commercialized transconjugate *Bt* products and the range of insect pests controlled by them are given in Table 3.1.

However, the conjugational approach to create novel *Bt* strains has certain limitations. Not all *Bt* toxin genes are located on transferable plasmids.

Table 3.1   Commercial transconjugate *Bt* products

Bt *recipient Strain*	*Conjugative Plasmid*	*Trade Name*	*Company*	*Target*
Btk	*Bt-aizawai*	Condor (EG 2348)	Ecogen	Soybean looper, velvet bean caterpillar, green clover-worm, gypsymoth and spruce budworm.
Btk	*Bt-aizawai*	Cutlass (EG 2371)	Ecogen	Beet armyworm cabbage looper, diamondback moth and cabbage web worm.
Bt-aizawai	*Btk*	Agree (GC-91)	Thermo Trilogy	Lepidoptera, especially diamondback moth.
Bt-aizawai	*Bt*	Design	Thermo Trilogy	Lepidoptera, especially diamondback moth.
Btk (HD-263)	*Bt-tenebrionis*	Foil* (EG 2424)	Ecogen	Colorado potato beetle, European corn borer armyworm and looper.

* Discontinued

Second, the toxin protein with useful insecticidal activity may be synthesized at low amount. Plasmid incompatibility could also be a problem. Conjugation is not easily controllable in the laboratory and therefore it limits the number of toxins (generally only 2-3) that can be present in the final strain.

2.2   Electroporation

Electroporation technique offers another approach of genetic transformation of *Bt* cells. Through a short electrical discharge, a temporary and reversible breakdown of the cell membrane is achieved, which allows high molecular weight substances such as DNA to enter the cell..This enables the introduction of cloned toxic genes back into various *Bt* strains. For example, a *Bt-tenebrionis* gene was introduced into *Bt-israelensis*, resulting in a dipteran-active strain with additional activity against *Pieris brassicae*, a property which neither parent strain possessed (Crickmore et al., 1990).

2.3   Transduction

Transduction is the transfer of bacterial DNA between cells (intra- or inter-serotype) via transducing phage particles, a technique that is useful for both gene mapping and the production of recombinant strains. Kalman et al. (1995) used a two step procedure to place a *cry1Ca* gene from *Bt-aizawai* into the chromosomes of two *Bt-kurstaki* strains, which contained multiple *cry* genes. In the first step, an integration vector was used to place *cry1Ca* gene into the chromosome of *Bt-kurstaki* HD-73 via electroporation. In the second step, a generalized transducing bacteriophage was used to transfer the integrated *cry1Ca* gene from *Btk* HD-73 to two other *Btk* strains, thus producing strains with a broader insecticidal spectrum.

2.4   Classical Mutation

High-energy radiation has proved to be one of the most effective modes of mutagenizing genes. Novo Nordisk used classical mutation to improve *Bt-tenebrionis*. This strain produces a bigger crystal, which is directly correlated with enhanced field activity. The commercial product Novodor® (Abbott) is based on a mutant *Bt-tenebrionis* strain NB176 containing two or three copies of Cry3A and is obtained by gamma irradiation of *Bt-tenebrionis* strain NB125. It produces unusually large rhomboid crystals composed primarily of Cry3A toxin (Gurtler and Petersen, 1991).

Although these genetic manipulation techniques have the capability to increase potency and control specificity, they are limited by several factors due to which it is not possible to construct a strain containing only the desired δ-endotoxins genes. The tools of genetic engineering can be used to overcome all of these limitations.

3   Recombinant DNA Technology

The possibility of transferring specific genes from one cell to another could give rise to a wide array of new products with desired characteristics. The discovery of restriction enzymes as molecular "scissors" to cut DNA into reproducible fragments at specific sites has been used to create new molecules by splicing or recombining two different pieces of DNA. Restriction enzymes can be used to cut specific sequences of DNA (genes) and to cut open a bacterium's plasmid (cloning vector). The cut ends of the gene and the cut ends of the plasmid are chemically "sticky", so they attach to each other (recombine), to form a new circular plasmid containing the new gene (Figure 3.1). The recombinant plasmid carries genetic instructions for the production of a new protein and when inserted into a bacterium, it produces this new protein.

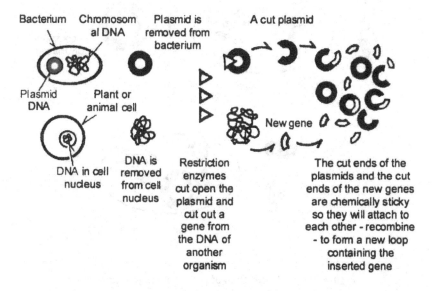

Figure 3.1   Recombinant DNA technology.

Recombinant DNA technology provides the tools for developing safe, efficient and cost-effective microbial control agents. It is now possible to combine the best traits of several different organisms into a single strain, expressing δ-endotoxins that exhibit enhanced insecticidal activity, longer residual activity and broader host range. Modified Cry proteins engineered for improved production or toxicity can also now be readily used as active ingredients. Conjugation and transduction have been used to transfer recombinant plasmids from donor to a recipient *Bt*; however the preferred method of gene transfer employs the use of electoporation, for which numerous protocols are available (Baum et al., 1999).

A number of convenient shuttle vectors functional in *E. coli* and *Bacillus* species have been constructed using replication origins from resident *Bt* plasmid. These shuttle vectors, which exhibit good segregational stability, are employed to introduce new toxin *Bt* genes to *Bt* strains (Figure 3.2).

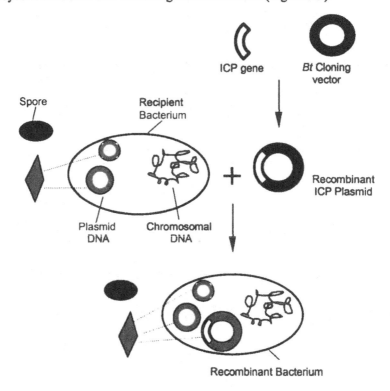

Figure 3.2   Construction of genetically improved *Bt* strain.

Alternatively, integrational vectors have been used to insert *cry* genes by homologous recombination (recombination between homologous segments on different DNAs) into resident plasmids or the chromosome.

A new *Bt*- based plasmid vector system enabled the introduction of a *cry3A* gene into *Btk* (HD-119) without affecting the expression levels of the native *cry* genes (Gamel and Piot, 1992). Similarly, cloning of a *cry1Ac* gene from *Btk* into a *Bt-aizawai*, resulted in enhanced host spectrum. The original *Bt-aizawai* showed good activity against *Spodoptera exigua*. The recombinant product showed additional activity against a range of other species, including *Helicoverpa zea, Heliothis virescens, Trichoplusia ni* and *Plutella xylostella* (Baum et al., 1990) (Figure 3.3).

In many instances, it has been found that introduced plasmid vectors carrying isolated *cry* genes are unstable in *Bt*. Often, in the absence of selection pressure, all or a portion of these plasmids are lost. The problem of plasmid instability of introduced genes can be overcome by integrating cloned *cry* genes into the chromosomal DNA of the host cell. A *cry1C* gene was introduced into the chromosomal DNA of *Bt-kurstaki* to broaden its host range (Glick and Pasternak, 1998). The transformed *Bt-kurstaki* strain showed a six-fold increase in its ability to kill *Spodoptera exigua* larvae.

Development of novel *Bt*-based cloning vectors, such as Ecogen's proprietary site-specific recombination (SSR) system, has made it possible to construct improved *Bt* strains for use as microbial insecticide. This system could selectively delete ancillary or foreign DNA elements (e.g. antibiotic resistance genes) from recombinant *cry* plasmids after their introduction into a *Bt* host. This has been used in the construction of a recombinant Cry3-overproducing strain EG 7673 that became the active ingredient of the recombinant product Raven®. The strain contains two coleopteran active genes, *cry3A* and *cry3Bb*, in addition to *cry1Ac*. Apparently, the strain's overproduction of Cry3 protein

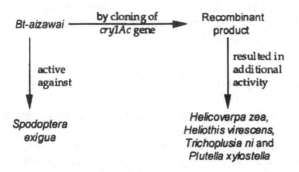

Figure 3.3   Recombinant *Bt-aizawai* cloned with *cry1Ac* gene from *Btk* shows an enhanced host spectrum.

allows for more cost-effective use of the product for the control of Colorado potato beetle (*Leptinotarsa decemlineata*) larvae (Baum et al., 1990).

Fusion of the coding portions of the active regions of two different toxin genes is another way of generating a novel protein with extended toxicity. *In vivo* recombination was used to produce hybrid genes constructed from truncated *cry1Aa* and *cry1Ac* genes, and two hybrid gene products produced in this way acquired an entirely new activity against *Spodoptera littoralis* (Cramori et al., 1991).

Thus, advances in recombinant DNA technology, which facilitated the cloning of toxin genes and their expression in plants and various other organisms, have increased the feasibility of using *Bt* for insect control. As a result of the enormous potential that recombinant DNA technology provides for advances in plant protection and other future insect control strategies based on *Bt*, almost every major chemical company, and many emerging biotechnology companies worldwide, are investing heavily in this area of research.

3.1   Bt *Transgenic Microorganisms*

The isolation of δ-endotoxin has enabled scientists to characterize and clone the genes encoding these molecules. Each δ-endotoxin is encoded by a single gene, therefore it can be easily transferred to other *Bt*s or to other microorganisms, such as, bacteria, algae, fungi, virus etc. to create more stable/or compatible agents for the toxin delivery. Genetic engineering technology allows the use of microorganisms that multiply on (epiphyte) or in (endophyte) plants to continuously produce insecticidal protein at the site of feeding. The primary rationale for using endophytic or epiphytic bacteria as hosts is to prolong the persistence of Cry proteins in the field by using a host that can propagate itself at the site of feeding and continue to produce crystal protein. By expressing engineered *cry* genes in the transgenic hosts, their expression can be boosted to higher levels than that in wild-type strains and hence can overcome broad-spectrum resistance (Koziel et al., 1993b).

3.1.1   *Hosts for epiphytic delivery of* Bt *genes*

In the early 1980s, Monsanto Company developed a recombinant, plant-colonizing *Pseudomonas* for delivery of *Bt* genes, with the objective of improving residual activity and efficacy of *Bt* proteins. Mycogen Corporation further developed the concept to the first genetically engineered bacterial insecticide MVP® active against lepidopterous insects. They introduced cloned *Bt cry1Ac* toxin gene into a microbial host, the root colonizing gram-negative bacteria strains of *Pseudomonas fluorescens*. The recombinant *P. fluorescens* cells that were subsequently killed by a proprietary chemical treatment (CellCap® technology), cross-linked the bacterial cell wall to yield a non-viable encapsulated bacterium surrounding the crystal protein (Figure 3.4).

A number of commercial products have been developed based on this process, including MVP®II, Mattch® and M-Peril® that are sprayed on the crop like other *Bts* (Table 3.2). The apparent advantage of these strains over conventional *Bts* is that pseudomonad cell protects the *Bt* protein from environmental degradation, thus providing longer residual activity in the field. Mycogen (1998) reported that MVP®II provided more persistent, long lasting control of caterpillar pests than conventional Bt products. The exact mechanism by which the cell protects the biotoxin is unknown, but the fixed cell wall provides a mechanical barrier and may have the ability to impede a variety of denaturing effects, including inactivation by sunlight (Gaertner et al., 1993). A recombinant *Bacillus megaterium* expressing a *cry1Aa* gene from *Btk* HD-1 also reportedly persisted for more than 28d under field conditions, whereas *Btk* disappeared within 4 days (Bora et al., 1994).

Stock et al. (1990) used cloning and conjugation technique to engineer *cry1Ac* gene from *Bt-kurstaki* HD-1 into plant colonizing bacterium *Pseudomonas cepacia* 526. The transconjugant *P. cepacia* cells produced only a truncated (78 kDa) Cry protein, which protect axenically grown tobacco plants from infestation by tobacco's hornworm.

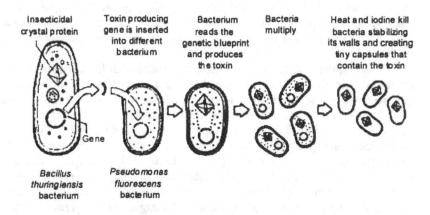

Figure 3.4   A *Bt cry* gene is transplanted into another bacterium *Pseudomonas fluorescens*. These bacteria, when killed, form tiny capsules containing the toxin protein (Source: Mycogen Corporation).

3.1.2   Hosts for endophytic delivery of Bt genes

An endophyte has been used to deliver δ-endotoxin to soil inhabiting insects. The *cry3A* gene from *Bt-tenebrionis* was introduced into *Rhizobium meliloti* and *R. leguminosarum* BV *vicaea* to protect alfalfa and pea crops from the feeding damage of two coleopteran insects, larvae of clover root curculio (*Sitona hispidulus*) and pea leaf weevil (*Sitona lineatus*) respectively (Bezdicek et al., 1991) (Table 3.2).

Scientists at Crop Genetics International transformed the *cryIAc* gene from *Bt-kurstaki* HD-73 into the gram-positive endophytic bacterium *Clavibacter xyli* subsp. *cynodontis* (Cxc), for the control of European corn borer. The transgenic Cxc, InCide® infects the internal tissues of growing plants and during growth produces *Bt* toxin. As the corn borer larvae feed in the stem, the insects ingest the endophyte and the *Bt* (Lampel et al., 1994).

3.1.3   Hosts for Bt genes for mosquito control

One of the major factors limiting the duration of mosquito control following the application of *Bt-israelensis* is the rapid sedimentation of spores and insecticidal crystals as the most mosquito larvae feed at or near the water surface. A potential approach is to engineer microorganism living in the upper layer of aquatic habitat to synthesize toxic protein. For example, *cry4B* toxin gene from *Bti* was cloned into the broad host range plasmid pRK248 and expressed it in *Caulobacter crescentus* CB15, a motile ubiquitous bacteria. The recombinant *Caulobacter* can provide the potential for prolonged control (Thanabalu et al., 1992).

Mosquitocidal toxin genes have been also shuttled between *Bacillus sphaericus* and *Bt-israelensis* to extend the host range to increase the toxicity and persistence and to help delay the appearance of resistant mosquitoes. Indeed, as the receptors for *B. sphaericus* and *Bti* toxins are different, placement of *Bti* toxin genes into *B. sphaericus* recipient cells, would give benefits of both worlds, the wider spectrum of *Bti* in a bacterium capable of surviving, persisting and recycling in the environment (Singer, 1990). When *B. sphaericus* strain 2362 was transformed with replicative plasmids containing the *cryIIA* gene, the transformants obtained showed various segregational or/and structural stabilities, some were moderately toxic to *Aedes aegypti* larvae (Trisrisook et al., 1990). Similarly, recombinants expressing *cry4B* and *cytIA* genes, independently or in combination, in *B. sphaericus* strain 2362, were more toxic to *Ae. aegypti* than the parental strain (Bar et al., 1991).

A novel approach for expressing *Bti cryIIA* gene in *B. sphaericus* by homologous recombination has been described (Poncet et al., 1997). The recombinant strain produced a large amount of Cry11A during the sporulation

Table 3.2   *Bt* Transgenic microorganisms

Transformed Organism	Gene Source	Process	Trade Name	Company/ Reference	Target
Pseudomonas fluorescens	*cryIAc* (*Btk*)	rDNA / CellCap	MVP II	Mycogen	Diamondback moth, loopers, tobacco budworm and cotton bollworm
Bacillus Megaterium (Strain RS1)	*cryIAa* (Btk HD-1)	rDNA		Bora et al., (1994)	Cotton bollworms
Bacillus Megaterium	*cryIIA* (*Bti*)	rDNA		Donovan et al.,(1988)	Mosquito larvae
Clavibacter xyli cynodontis (MD 6PA)	*cry IAc* (*Btk* HD-73)	rDNA	InCide*	Lampel et al., (1994)	European corn borer
Rhizobium meliloti	*cry3A* (*Bt-tenebrionis*)	rDNA		Bezdicek et al., (1991)	Colorado potato beetle, clover root curculio and pea leaf beetle
Caulobacter Crescentus CB 15	*cry4B* (*Bti*)	Shuttle vector		Thanabalu et al., (1992)	Mosquito larvae at or near the water surface
Bacillus sphaericus	*cryIIA* (*Bti*)	Homologous Recombi-nation		Poncet et al., (1997)	*Aedes aegypti* and *Culex quinqui-faciatus*
B. sphaericus (2362)	*cry4B*, *cytIA* (*Bti*)	rDNA		Bar et al., (1991)	*Aedes aegypti*
B. sphaericus (2362)	*cryIIA* (*Bti*)	rDNA		Trisrishook et al.. (1990)	*Aedes aegypti*
Pseudomonas fluorescens	*cryIAc*, *cryIC* (*Bt-aizawai*)	rDNA./ CellCap	Mattch	Mycogen	Broad spectrum of caterpillar pests, such as beet army-worm, diamondback moth, corn earworm etc.

Table 3.2 (continued)

Transformed Organism	Gene Source	Process	Trade Name	Company/ Reference	Target
Pseudomonas fluorescens	Btk	rDNA/ CellCap	M-Peril	Mycogen	European corn borer, South western corn borer and fall armyworm
Bt-kurstaki	cry1Ac (2), cry3A and cry3Bb	rDNA	Raven (EG 7673)	Ecogen	Lepidoptera and coleoptera insects on potatoes, tomatoes and eggplant
Btk	cry1Aa, cry1Ac (2), cry2A, and cry 1F-1Ac	rDNA/ Protein Engineering	Lepinox (EG 7826)	Ecogen	Fall and beet armyworm in sweet corn, turf and other raw crops
Btk	cry1Ac (3), cry1C and cry2A	rDNA/ Protein Engineering	Crymax (EG 7841)	Ecogen	Broad spectrum activity against caterpillars on fruits and vegetables

* Not commercialized

process. The recombinant strain produced more binary toxin than did the parental strain. Synthesis of the Cry11A toxin conferred toxicity to the recombinant strain against *Aedes aegypti* larvae, for which the parental strain was not toxic. Interestingly, the level of larvicidal activity of recombinant strain against *An. stephensi* was as high as that of *Bti*. The toxicities of parental and recombinant *B. sphaericus* strains against *Culex quinquefasciatus* were similar, but the recombinant strain killed the larvae more rapidly.

The *Bt-israelensis* toxin gene *cry11A* has also been cloned into *B. megaterium* and the transformed bacterial cells were highly toxic to mosquito larvae (Donovan et al., 1988). Similarly, the *cry4A* gene of *Bt-israelensis* was introduced into various unicellular cynobacteria with the intent of providing a more accessible source of toxin for filter feeding dipteran larvae (Angsuthana-sombat and Panyim, 1989) (Table 3.2).

3.2 *Protein Engineering*

Large scale screening of field-isolated *Bt* strains has been quite successful over the years in identifying new *Bt* strains with increased levels of insecticidal activity as well as activities against a wide range of insect pests. Although new crystal protein-encoding genes are being identified in these strains at an ever-increasing rate, it is becoming apparent that highly active toxins, specific for certain major pest species, may not be readily found in nature. Today there exists a strong movement toward protein design and engineering to help overcome some of the inadequacies of the crystal toxins. This protein engineering approach could also be effectively used as a model system in strain improvement and development of highly active and broader pest specificity toxins (Powell et al., 1995).

Most of the current emphasis on engineering of the crystal protein is aimed at characterizing their mode of action, that is, how the proteins exert their effects at the molecular level. The possibilities for engineering novel *Bt* crystal proteins with higher activities and wider host spectra are just beginning to be explored. Ecogen (1998) reported technology to improve the insecticidal capabilities of the *Bt* proteins through its work on the mechanisms involved in the insecticidal activity. This technology, collectively referred to as mechanism assisted insecticidal design, permits moving beyond naturally occurring proteins to identify new and commercially valuable proteins. Lepinox®, a product based on recombinant DNA and protein engineering technology to create novel insecticidal proteins, has been commercialized. It has increased activity against several species of armyworms, while maintaining the favorable environmental and safety aspects of a naturally occurring *Bt*. Similarly, another product Crymax®, that contains a combination of three protein toxins Cry1Ac, a modified Cry1C and Cry2A, provides a high level of potency and spectrum of activity against caterpillar pests (Table 3.2).

Rajamohan et al. (1996) reported that by using site-directed mutagenesis techniques, there is a significant enhancement in toxin-receptor contact and subsequently improved binding affinity and potency of Cry1Ab toxin. Wu and Aronson (1992) observed that a mutation in helix $\alpha5$ of domain I of Cry1Ac, caused a two-fold increase in toxicity against *M. sexta*, that was found to be correlated with the rate of irreversible binding.

4   Insect-Tolerant Transgenic Crop Plants

The use of *Bt* for crop protection has evolved rapidly, from the direct application of the spores and crystal toxins of this bacterium as biopesticide, to the development of transgenic plants expressing cloned toxin genes. Genetic engineering has created transgenic variety of many crop plants that express *Bt-*

toxins (Huang et al., 1999: Schuler et al., 1998). Although the potential exists to transfer most of the insecticidal principles found in diverse organisms, the current work has focused on transfer of insecticidal genes of bacterial origins, in particular from *Bacillus thuringiensis* to various crop plants. The transgenic plants armed with *Bt*-toxins are defended against some of the most notorious pests, which reduce the need for insecticidal sprays.

The transgenic plants provide season long protection, independent of weather conditions, effective control of burrowing insects and others that feed at sites difficult to reach with sprays and control at all the stages of insect development. The important feature of such a system is that only insects eating crops are exposed to the toxins. Because *Bt* is not toxic to arthropod natural enemies, opportunities for biological control are enhanced and the secondary pest outbreaks often caused by conventional pesticides are avoided. Such transgenic crops provide farmers an attractive alternative over traditional application of chemical pesticides and a means of controlling serious insect pests, not easily controlled by current chemical pesticides. The use of transgenic plants could also overcome some of the stability problems associated with conventional *Bt* applications. Thus, the new technology could yield enormous benefits for food production and environment quality worldwide.

Bt strains contain great diversity of δ-endotoxins encoding genes and have proved to be a remarkable source of insecticide principles to be used in transgenic plants. *Bt*-insecticidal proteins have been successfully expressed in a wide range of transgenic plants, including cotton, corn, tomatoes, potatoes, rice and soybean. Their expression have been achieved at levels that provide economic control of lepidopterous pests, including cotton bollworms, tobacco budworm, pink bollworm and the European cornborer. This has introduced a new dimension, in the utility of *Bt*-toxins as a crop protectant in transgenic plants. Several companies including Monsanto and Novartis have received registration for transgenic cotton and corn seeds by the EPA. The first generation of insect resistant plants, such as transgenic corn, cotton and potatoes were grown on a large scale in the United States during 1996.

4.1   Dicotyledonous Plants

The *Bt* genes *cryIAa*, *cryIAb* or *cryIAc* isolated from lepidopteran active *Btk*, have been introduced into tobacco, tomato or potato. The toxin genes were introduced by *Agrobacterium*-mediated transformation (Figure 3.5). *Agrobecterium tumafaciens* is a plant pathogen causing tumorous crowngalls on infected dicotyledonous plants. *Agrobacterium*-mediated gene transfer has been a well established method to introduce foreign genes into plants, albeit restricted to dicotyledonous plants and their expression conferred some degree of protection against tobacco pests (*Manduca sexta*), tomato pests (*Heliothis*

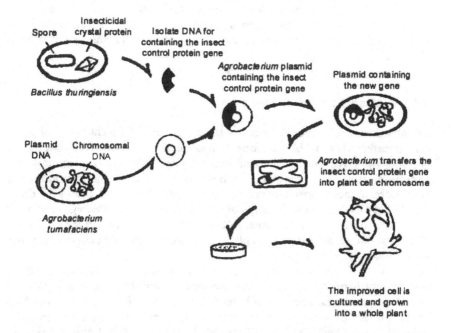

Figure 3.5   Cotton plant resistant to insect pests. Each cell contains the insecticidal protein gene, providing the cotton plant with the ability to ward off attack by caterpillar pests (Courtesy Monsanto Company).

virescens and *Helicoverpa zea*) and potato pests (*Phthorimaeq operculella*). Perlak et al. (1990) reported on the performance of transgenic cotton plants (var. Coker 312) containing *crylAb* or *crylAc* with effective control of cotton pests such as tobacco budworm (*Heliothis virescens*), pink bollworm (*Pectinophora gossypiella*) and moderate population levels of cotton bollworm (*H. zea*).

The development of cotton cultivars with transgenes usually involved two distinct phases. The first phase involved selection of transformation events that express the δ-endotoxin protein at the desired level, without any major negative effects on agronomic and fiber properties. The second phase involved hybridization of the selected transformed plant with elite germ plasm, followed by selection and evaluation of progeny to determine the expression of the δ-endotoxin and the agronomic and fiber properties of the selected lines (Jenkins, 1999). A modified δ-endotoxin-encoding gene *cry3A*, isolated from coleopteran active *Bt-tenebrionis* was transformed into potato plants conferring protection from damage by the Colorado potato beetle (*Leptinotarsa decemlineata*) at high levels of field infestations (Adang et al., 1993; Perlak et al., 1993). These initial accomplishments were restricted to dicotyledonous plants.

 Bt cotton has been released commercially in the USA, Australia and South Africa. *Bt* potato has been approved for commercialization in USA, Canada and Japan (Schuler et al., 1998). The commercial transgenic lines of cotton and potato are given in Table 3.3.

4.2   Monocotyledonous Plants

Transformation of monocotyledonous plants, is based on technology of direct gene transfer, where DNA transformation is mediated by non-biological means. Polyethylene glycol-mediated transformation, electroporation or microprojectile bombardment has been developed as possible approaches. The microprojectile bombardment looks especially promising, primarily because it allows the direct delivery of DNA into a wide range of plant cells, obviating the regeneration of plants from transferred protoplasts. Koziel et al. (1993a) reported the expression of *Bt*-genes in monocotyledonous plants by transforming elite cultivars of maize with a truncated *cry1Ab* gene.

 Ciba Seeds (now Novartis Seeds) and Mycogen Seeds introduced the first *Bt* corn hybrids in 1996. These transgenic corn plants provided excellent protection against European corn borer (*Ostrinia nubilalis*) even at a very high insect larvae pressure. Several seed companies have incorporated this technology into their best inbred lines. Seed companies select elite hybrids for the *Bt* transformation in order to retain important agronomic qualities for yield, harvestability and disease resistance. Various commercialized corn lines contain improved expression of one of three Cry protein genes active against European corn borer, i.e. Cry1Ab, Cry1Ac or Cry9C. Most of the *Bt* corn hybrids, produce only the Cry1Ab protein; a few produce the Cry1Ac protein or the Cry9C protein.

Table 3.3   Commercialized transgenic *Bt* cotton and potato

Plants	Gene	Trade Name	Company	Target
Cotton	*cry1Ac*	Bollgard	Monsanto	Bollworm and budworm
Potato	*cry3A*	NewLeaf	Monsanto/ NatureMark	Colorado potato beetle
Potato	*cry3A +* *Potato Leaf Roll* *Virus resistance* *gene*	NewLeaf Plus	Monsanto/ NatureMark	Colorado potato beetle and potato leaf roll virus

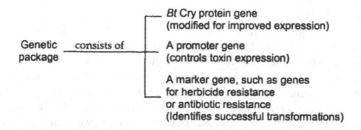

Figure 3.6   Genetic package components

The genetic package inserted into corn consists of three primary components: a *Bt* Cry protein gene, a promoter gene that controls where, when and how much of the toxin is expressed and a marker gene that allows identification of successful transformations (Figure 3.6). Current examples of markers include genes for herbicide-tolerance or antibiotic-resistance.

Successful transformations, called "events," vary in the components of the genetic package and where this DNA is inserted into the corn DNA. The insertion site may affect *Bt* production and could affect other plant functions (Figure 3.7). The level of European corn borer (ECB) control against late-season ECB infestations differs between *Bt* events. Under heavy ECB pressure, events BT11 and MON810 provide a higher level of control than event 176. The difference could be explained by the fact that event 176 hybrids produce *Bt* protein only in green tissues and pollen, whereas BT11 and MON810 events produce it throughout the plant. Because some hatching larvae initially colonize ears to feed on silks and developing kernels, these larvae may survive on

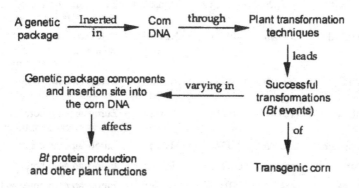

Figure 3.7   Construction of transgenic *Bt* corn

event 176 and may tunnel later into stalks and ear shanks (Ostlie and Hutchinson, 1997).

The transgenic *Bt* corn has been commercially released in the USA, Canada, Argentina, Japan and the European community (Schuler et al., 1998). The various commercialized *Bt* corn lines are given in Table 3.4.

A modified *cryIAb* gene has been inserted into rice using micro-projectile bombardment technique to enhance *cryIAb* gene expression. The level of expression of the modified gene in transgenic rice was 0.05% of total soluble leaf protein. The plants were significantly resistant to two lepidopteran rice pests, leaf folder (*Cnaphalocrosis medinalis*) and stemborer (*Chilo suppressalis*) (Fujimoto et. al., 1993).

4.3   Second Generation Insect Resistant Plants

The second generations of insect resistant plants are under development. These include both *Bt* and non-*Bt* proteins with novel modes of action and different spectra of activity against insect pests. For example, proteins such as cholesterol oxidases and members of VIPs (vegetative insecticidal proteins) family represent the second generation of insecticidal trans-genes that will complement the novel *Bt* δ-endotoxins. A maize plant has been transformed which expresses

Table 3.4   Commercialized transgenic *Bt* corn

Trade Name	Gene	Event	Company	Target
YieldGard	*cryIAb*	MON810	Monsanto	European corn borer, corn earworm, fall armyworm
YieldGard	*cryIAb*	BT11	Novartis Seeds	European corn borer, southwestern corn borer, corn ear worm and fall army worm
KnockOut	*cryI*Ab	176	Novartis Seeds	European corn borer
Attribute	*cryIAb*	BT11	Novartis Seeds	European corn borer and southwestern corn borer
NatureGard	*cryIAb*	176	Mycogen	European corn borer
Bt-Xtra	*cryIAc*	DBT418	Dekalb	European corn borer
StarLink	*cry9C*	CBH351	AgrEvo	European corn borer and black cutworm

a VIP and a Bt δ-endotoxin, by genetically engineering the plant to contain and express all the genes necessary (Warren et al., 1999). These can control economically important insect pests such as northern and western corn rootworm and bollweevil, which are not effectively controlled by any known *Bt* δ-endotoxin. In an effort to enhance the *Bt* δ-endotoxin gene expression in second generation plants, the genes with sequence modification and/or new promoters including tissue-specific promoters are being used (Estrusch et al., 1997). Biotech companies are also working to develop *Bt* crops with "stacked" or "pyramided" genes that can produce two different *Bt* toxins with two different modes of action This will reduce the likelihood of pest resistance and will lead to a relaxation in refuge requirements and other regulatory factors (Renner, 1999).

5   Concluding Remarks

The recent elucidation of the X-ray crystal structures of *Bt* proteins opens new possibilities for protein engineering and design of *Bt* insecticides. Determination of the mode of action at the molecular level and the genetic basis for insect specificity should enable recombinant DNA technology to be used to expand the insect host range of *Bt* as well as increase its toxicity against insects. In the future, it is easy to imagine that crystal protein variants, engineered for improved toxicity, yield or stability by *in vitro* mutagenesis, may be used to steadily improve the performance of *Bt*-based insecticides. In addition, the construction of hybrid toxins with improved insecticidal activity will provide novel active ingredients suitable for commercial exploitation. It is anticipated that engineered forms of toxin proteins showing improved potency or yield, regardless of their host, will make *Bt* bioinsecticides a more attractive and practical alternative to synthetic chemical control agents.

Advances in understanding of protein structure and function are allowing scientists using protein-engineering technologies, to begin to design chimeric proteins. These chimeras are constructed by piecing together at the gene level discrete functional parts of the protein, called domains that can be swapped to change or increase their normal species activity spectrum. The long-term goal of protein engineering is construction of modular "smart proteins" that will target specific pests and, like *Bt*, not harm beneficial animals.

The use of transgenic plants expressing *Bt* δ-endotoxin provide a level of insect control not seen since the introduction of chemical insecticides decades ago and have the potential to greatly reduce the environmental and health costs associated with the use of conventional insecticides. Advances in tissue culture have been combined with improvements in transformation technology to increase transformation efficiencies particularly in case of corn and cotton. It is anticipated that second generation transgenic plants producing insecticidal proteins with either different modes of action or different targets or both options consisting of multiple genes into the same plant would reduce the possibility of resistance development.

References

1.   Adang, M. J., M. S. Brody, G. Cardineau, N. Eagan, R. T. Roush, C. T. Shewmaker, A. Jones, J. V. Oakes and K. E. McBride, 1993. "The reconstruction and expression of a *Bacillus thuringiensis cryIIIA* gene in protoplasts and potato plants", *Plant Molecular Biol.*, 21, 1131-1145.

2.   Angsuthanasombat, C. and S. Panyim, 1989. "Biosynthesis of 130 kilodalton mosquito larvicide in the cynobacterium *Agmenellum quadruplicatum* PR-6", *Appl. Environ. Microbiol.*, 55, 2428-2430.

3.   Bar, E., J. Lieman-Hurwitz, E. Rahamim, A. Keynan and N. Sandler, 1991. "Cloning and Expression of *Bacillus thuringiensis israelensis* δ-endotoxin DNA in *B. sphaericus*", *J. Invertebr. Pathol.*, 57, 149-158.

4.   Baum, J. A., T. B. Johnson and B. C. Carlton, 1999. "*Bacillus thuringiensis*: Natural and recombinant bioinsecticide products", in *Biopesticides: Use and Delivery*, F. R. Hall and J. J. Menn, eds., Humana Press Inc., Totowa, N.J., pp. 189-209.

5.   Baum, J. A., D. M. Coyle, M. P. Gilbert, C.S. Jany and C. Gawron-Burke, 1990. "Novel cloning vectors for *Bacillus thuringiensis*", *Appl. Environ. Microbiol.*, 56, 3420-3428.

6.   Bezdicek, D. F., M. A. Quinn and M. Kahn, 1991. "Genetically engineering *Bacillus thuringiensis* genes into R*hizobia* to control nodule feeding insects", *USDA CRIS Report* No. 9131200.

7.   Bora, R. S., M. G. Murty, R. Shenbargarathai and V. Sekar, 1994. "Introduction of Lepidopteran-specific insecticidal crystal protein gene of *Bacillus thuringiensis* subsp. *kurstaki* by conjugal transfer into a *Bacillus megaterium* strain that persists in the cotton phylloplane", *Appl. Environ. Microbiol.*, 60, 214-222.

8.   Bosch, D., B. Schipper, H. van der Kleij, R. A. de Maagd and W. J. Stiekeme, 1994. "Recombinant *Bacillus thuringiensis* crystal proteins with new properties: possibilities for resistance management", Bio/Technology, 12, 915-918.

9.   Cannon, R. J. C., 1996. "*Bacillus thuringiensis* use in Agriculture: A molecular perspective", *Biol. Rev.*, 71, 561-636.

10.   Caramori, T., A. M. Albertini and A. Galizzi, 1991. "*In vivo* generation of hybrids between two *Bacillus thuringiensis* insect-toxin-encoding genes", *Gene*, 98, 37-44.

11.   Carlton, B. C. and C. Gawron-Burke, 1993. "Genetic improvement of *Bacillus thuringiensis* for bioinsecticide development", in *Advanced Engineered Pesticides*, L. Kim, ed., Marcel Dekker, Inc., New York, pp. 43-61.

12. Crickmore, N., C. Nicholls, D. J. Earp, C. Hogman and D. J. Ellar, 1990. "The construction of *Bacillus thuringiensis* strains expressing novel entomocidal delta-endotoxin combinations", *Biochem. Jour.*, 270, 133-136.

13. Donovan, W. P., C. Dankocsik and M. P. Gilbert, 1988. "Molecular characterization of a gene encoding a 72-kilodalton mosquito-toxic crystal protein from *Bacillus thuringiensis* subsp. *israelensis*", *J. Bacteriol.*, 170, 4732-4738.

14. Estruch, J. J., N. B. Carozzi, N. Desai, N.B. Duck, G.W. Warren and M.G. Koziel, 1997. "Transgenic plants: an emerging approach to pest control", *Nature Biotechnol.*, 15, 137-141.

15. Ecogen Inc., 1998. "CRYMAX® Bioinsecticide", http://stak1.voicenet. com /~ecogen01/Pages/ crymax.html.

16. Fujimoto, H., K. Itoh, M. Yamamoto, J. Kyozuka and K. Shimamoto, 1993. "Insect resistant rice generated by introduction of a modified δ-endotoxin gene of *Bacillus thuringiensis*", *Bio/Technology*, 11, 1151-1155.

17. Gaertner, F. H., T. C. Quick and M. A. Thompson, 1993. "CellCap: an encapsulation system for insecticidal biotoxin proteins", in *Advanced Engineered Pesticides*, L. Kim, ed., Marcel Dekker, New York, NY, pp. 73-83.

18. Gamel, P. H. and J.-C. Piot, 1992. "Characterization and properties of a novel plasmid vector for *Bacillus thuringiensis* displaying compatibility with host plasmids", *Gene*, 120, 17-26.

19. Glick, B. R. and J. J. Pasternak, 1998. "Microbial insecticides", in *Molecular Biotechnology: Principles and Applications of Recombinant DNA*, 2nd edition, ASM Press, Washington, D. C., pp. 377-398.

20. Gurtler, H. and A. Petersen, 1991. "Mutants or variants of *Bacillus thuringiensis* producing high yields of delta endotoxin", Patent WO 91/07481, World Intellectual Property Organization.

21. Huang, F., R. A. Higgins and L. L. Buschman, 1999. "Transgenic *Bt*-plants, successes, challenges and strategies", *Pestology, special issue*, Febuary 1999, 2-29.

22. Jenkins, J. N., 1999. "Transgenic plants expressing toxins from Bacillus thuringiensis", in *Biopesticides: Use and Delivery*, F. R. Hall and J. J. Menn, eds., Humana Press, Totowa, NJ, pp. 211-232.

23. Kalman, S., K. L. Kiehne, N. Cooper, M. S. Reynoso and T. Yamamoto, 1995. "Enhanced production of insecticidal proteins in *Bacillus thuringiensis* strains carrying an additional protein gene in their chromosome", *Appl. Environ. Microbiol.*, 61, 3063-3068.

24. Koziel, M. G., G. L. Beland, C. Bowman, N. B. Carrozi, R. Crenshaw, L. Crossland, J. Dawson, N. Desai, M. Hill, S. Kadwell, K. Launis, K. Lewis, D. Maddox, K. McPherson, M. R. Meghji, E. Merlin, R. Rhodes, G. W. Warren, M. Wright and S. V. Evola, 1993a. "Field performance of elite transgenic maize plants expresssing an insecticidal protein derived from *Bacillus thuringiensis*", *Bio/Technology*, 11, 194-200.

25. Koziel, M. G., N. B. Corozzi, T. C. Currier, G. W. Warren and S. V. Evola, 1993b. "The insecticidal crystal proteins of *Bacillus thuringiensis*: Past, Present and future uses", *Biotech. Gent. Eng. Rev.*, 11,171-228.

26. Kumar, P. A., R. P. Sharma and V. S. Malik, 1996. "The Insecticidal protein of *Bacillus thuringiensis*", *Adv. Appl. Microbiol.*, 12, 1-43.

27. Lampel, J. S., G. L. Canter, M. B. Dimock, J. L. Kelly, J. J. Anderson, B. B. Uratani, J. S. Foulke, Jr. and J. T. Turner, 1994. "Integrative cloning, expression and stability of the *cry IA(c)* gene from *Bacillus thuringiensis* subsp. *kurstaki* in a recombinant strain of *Clavibacter xyli* subsp. *cynodontis"*, *Appl. Environ. Microbiol.*, 60, 501-508.

28. Mycogen Tipsheet - MVP®II, 1998. http://www.mycogen.com/graphic/ pest/ tipsheet/mvp/2213.htm

29. Ostlie, K., and W. Hutchinson, 1997. "*Bt* corn & European corn borer", University of Minnesota Extension Service, http://www.mes.umn.edu/ Documents/D/C/DC7055.html

30. Perlak, F. J., T. B. Stone, Y. M. Muskopf, L. J. Petersen, J. Wyman, S. Love, G. Reed, G. Biever and D. A. Fischhoff, 1993. "Genetically improved potatoes: protection from damage by Colorado potato beetles", *Plant. Mol. Biol.*, 22, 313-321.

31. Perlak, F. J., R. W. Dealton, T. A. Armstrong, R. l. Fuchs, S. R. Sims, J. T. Grreenplate and D. A. Fischhoff, 1990. "Insect-resistant cotton plants", *Bio/Technology*, 8, 939-943.

32. Peferoen, M., 1997. "Progress and prospects for field use of *Bt* genes in crops", *Trends Biotechnol.*, 15, 173-177.

33. Poncet, S., C. Bernard, E. Dervyn, J. Cayley, A. Klier and G. Rapoport, 1997. "Improvement of *Bacillus sphaericus* toxicity against dipteran larvae by integration via homologous recombination of the *cryllA* toxin gene from *Bacillus thuringiensis* subsp. *israelensis"*, *Appl. Environ. Microbiol.*, 63(11), 4413-4420.

34. Rajamohan, F., O. Alzate, J. A. Cotrill, A. Curtiss and D. H. Dean, 1996. "Protein engineering of *Bacillus thuringiensis* δ-endotoxin: mutations at domain II of Cryl Ab enhance receptor affinity and toxicity towards gypsy moth larvae", *Proc. Natl. Acad. Sci. USA*, 93, 14338-14343.

35. Renner, R., 1999. "Will Bt-based pest resistance management plans work?", *Environ. Sci. Technol.* 33(19), 410A-415A.

36. Schuler, T. H., G. M. Poppy, B. R. Kerry and I. Denholm, 1998."Insect-resistant transgenic plants", *TIBTECH*, 16, 168-175.

37. Singer, S., 1990. "Introduction to the study of *B. sphaericus* as a mosquito control agent", in *Bacterial Control* of *Mosquitoes* and *Blackflies*, H. de Barjac and D. J. Sutherland, eds., London, Unwin Hyman, pp.221-227.

38. Stock, C. A., T. J. McLoughlin, J. A. Klein and M. J. Adang, 1990. "Expression of *Bacillus thuringiensis* crystal protein gene in *Pseudomonas cepacia* 526", *Can. J. Microbiol.*, 36, 879-884.

39. Thanabalu, T., J. Hindley, S. Brenner, C. Oei and C. Berry, 1992. "Expression of the mosquitocidal toxins of *B. sphaericus* and *Bacillus thuringiensis* subsp. *israelensis* by recombinant *Caulobacter crescentus*, a vehicle for biological control of aquatic insect larvae, *Appl. Environ. Microbiol.*, 58, 905-910.

40. Trisrisook, M., S. Pantuwatana, A. Bhumiratana and W. Panbangred, 1990. "Molecular cloning of the 130-kilodalton mosquitocidal □□endotoxin gene of *Bacillus thuringiensis* subsp. *israelensis* in *B. sphaericus"*, *Appl. Environ. Microbiol.*, 56, 1710-1716.

41. Warren G. W., M. G. Koziel, M. A. Mullins, G. J. Nye, B. Carr, N. M. Desai, K. Kostichka, N. B. Duck and J.. J. Estruch (Novartis Corporation), 1999. "Stably transformed plants comprising novel insecticidal proteins", U.S. Patent Number 5,990,383 (November 23, 1999).

42. Wu, D., and A. I. Aronson, 1992. "Localized mutagenesis defines regions of the *Bacillus thuringiensis* δ-endotoxin involved in toxicity and specificity", *J. Biol. Chem.*, 267, 2311-2317.

4

Formulation of Bacterial Insecticides

1   Introduction

The purpose of formulation of a microbial system is to produce a product that optimizes the effective and economical use of a toxin, within a particular environment. *Bacillus thuringiensis subsp. israelensis (Bti)* formulation, a mosquito biolarvicide, for example, is used in an aquatic environment. On the other hand, *Bacillus thuringiensis subsp. kurstaki (Btk)* formulation, a product for control of agricultural insect pests, is applied through foliar application. In order to be effective, both of these products should be easily accessible to their respective target larvae and should be reasonably stable under field conditions. Thus, even though the nature of insecticidal crystal proteins (ICPs) and spores are similar in both cases, the final product has to be applied to different environments, with specific requisites for accessibility and stability.

There have been a number of recent reviews on the formulation and application of bacterial and viral insecticides (Burges et al., 1998; Devisetty et al., 1998; Shieh, 1998; and Cibulsky et al., 1993). Various aspects and attributes of ingestible microbial formulations targeting for maximum effectiveness with specific reference to *Bt and Bs*, are discussed in this chapter.

2   Characteristics of Microbial Insecticide Formulations

In general, formulation processes and specifications of microbial systems are quite similar to that of chemical pesticides. However, microbials such as bacteria or viruses differ from chemical pesticides in several aspects as follows:

2.1   Particulate Characteristics

Insect pathogens are particulate materials composed of high molecular weight proteins. Unlike synthetic pesticide molecules, the toxin particulates are insoluble in aqueous and organic solvents. In fact, any effort to dissolve particulate toxins causes loss of toxicity. These are living entities with an organized cellular structure entrapped in a generally hydrophobic membrane. This limits the options available for formulating microbial pesticides to wettable powders, suspension concentrates and water dispersible granules. Surface bonded granulars and briquette formulations are also possible for specific applications. However the insoluble nature of the active ingredient does not permit formulations such as emulsifiable concentrates and water based concentrated emulsions.

2.2   Ingestion of Particulate Toxin by Target Larvae

Microbial agents such as bacteria, viruses etc. must be ingested by target pests in order to be toxic. Thus, palatability of biopesticide formulations assumes significance in terms of ingestible particle size, as well as role of feeding stimulants in enhancing ingestion rate.

2.3   Stability Characteristics

Microbial systems are generally more sensitive to the surrounding environment than most chemicals. Thus, the stability of microbial systems formulations requires special care for maintaining biological activity. Some of the factors that affect the stability and bioactivity of the formulations include pH, choice of surfactants and high temperature. The higher pH level often results in malodor formation and bacterial growth. Stabilization of *Bt* in aqueous vehicles is much more complex as several interacting factors such as hydrolysis, microbial degradation, oxidation, protein aggregation can adversely affect the stability. Similarly, use of anionic surfactants in an aqueous formulation may often denature the protein structure of toxin. Prolonged exposure to high temperature during the production process and subsequent storage may also result in loss of insecticidal activity.

2.4   Bioassay as Analytical Mode

A formulation active ingredient is analyzed by bioassay against species of pests. Alternative chemical methods, such as direct measurement of protein or immunoassay of δ-endotoxin in *Bt* formulations are valid only if extensive correlation studies are established between the bioactivity and chemical quantification.

3   Formulation Attributes Having Impact on Efficacy of Bacterial Insecticides

Insect pests develop in a multitude of ecological niches. Various factors need to be taken into account to develop a bacterial or viral formulation, including biology, susceptibility and feeding behavior of target pest(s), and information on type and nature of their breeding habitat. Some of the formulations can be tailor-made for a single site, pest or use-pattern. Different attributes that may lead to optimization of formulation efficacy are discussed here.

3.1   Particle Size Influences Placement and Ingestibility of Toxins

The particle size of the suspended formulations can exert considerable influence on application and efficacy. It is known that the settling rate of a formulation corresponds directly to its mean particle size. Thus, a microbial formulation for mosquito control that enables flotation of toxic constituents near the surface over a sustained period of time would enhance residual toxicity against surface feeding mosquito larvae.

It has been observed that a flowable formulation with a smaller mean particle size, settled more slowly than wettable powders and provided prolonged control. Khetan et al. (1996) studied the relative efficacies of two commercial formulations of *Bti* in laboratory and field against *Anopheline* larvae. The formulations compared were a wettable powder (Bactoculicide) of a higher potency (5400 IU/mg) and an aqueous suspension (Aquabac®) (1200 IU/mg). Application of an equal amount of both the formulations in the laboratory test resulted in higher mortality for powder formulation consistent with its relative higher potency. On the other hand, aqueous suspension formulation was found more effective in the field studies, when equal amounts of the two formulations

Table 4.1 Comparative efficacy of powder and aqueous suspension formulations of *Bti* in the field studies (Khetan et al., 1996)

Formulations of Bti	Potency[a] IU/mg	Reduction in Anopheline Larvae Population (%)			
		Days 1	3	7	14
Wettable Powder (Bactoculicide®)	5400	100	100	79	
Aqueous suspension (Aquabac®)	1200	100	100	84	47

(a) IU = International Units

were applied (Table 4.1). Investigations revealed that a fine particle sized aqueous suspension formulation of *Bti* out performed a higher strength powder formulation for control of aquatic surface feeding mosquito larvae.

The ingestion of particles by mosquito larvae is also governed mainly by size. Generally, mosquito larvae ingested particles between >0.5 µm and <50µm in diameter. In an X-ray diffraction analysis for particles ingested and retained by larvae of *Aedes triseriatus*, it has been observed that 1st and 2nd instar larvae ingested particles of <2 µm, 3rd instars ingested mostly particles in 2-10µm range, whereas, 4th instars ingested mainly particles in the range of 10-25µm. Only a small proportion of smaller particles (<2µm) was ingested by 3rd and 4th instar larvae (Table 4.2).

Skovmand et al. (1997) studied the influence of particle size on toxicity by conducting bioassays on 2nd and 4th instar larvae of *Ae. aegypti* employing several commercial *Bt-israelensis* formulations. Their findings revealed that smaller the median particle size, steeper the slope of the concentration-mortality curve, i.e. LC_{50} value becomes lower.

Table 4.2 Particles ingested and retained by larvae of *Aedes triseriatus* as determined by X-ray diffraction analysis

Aedes triseriatus larvae stage	Particle size ingested	
1st and 2nd instar	<2 µm	
3rd instar	2 – 10 µm	{A small proportion of smaller particles (<2 µm) was also ingested }
4th instar	10 – 25 µm (mainly)	

3.2 Feeding Stimulants Enhance Palatability and Anti-Feedants Reduce Efficacy of Bt Products

3.2.1 Feeding stimulants as adjuvants

The *Bt*-based bioinsecticides are gut toxins and must be ingested by the target larvae in order to effect mortality. Their toxic effects can be enhanced, if the feeding rate of the pests could be increased. To this end, use of a feeding stimulant or phago-stimulant would increase efficacy by increasing the concentration and rate of toxins ingested.

Lepidopteran larvae are reported to have a reduced feeding rate on foliage or artificial diet treated with *Btk*, or they avoid feeding in favor of untreated foliage. The feeding behaviour of tobacco budworm (*Heliothis virescens*) on artificial diet treated with *Btk* formulation was reported to result in feeding avoidance (Gould et al.,1991) (Figure 4.1). Similar observation of reduced feeding on Btk treated feed was reported with gypsymoth (*Lymantria dispar*) larvae. However, addition of nutrient based feeding stimulants (consisting of proteins, lipids and carbohydrates) in the formulation increased the initial dose ingested by the larvae, killing them and thus preventing recovery and feeding resumption (Farrar and Ridgway, 1995) (Figure 4.2).

Similarly, in mosquito larvae, Rashed and Mulla (1989) observed that dried yeast, wheat flour, fish meal and dried blood were ingested at significantly faster rates than inert particles such as kaolin, talc, chalk and charcoal. Yeast contains a number of phago-stimulants, which stimulate feeding activity of mosquito larvae and it increases ingestion rate in several species of mosquitoes too.

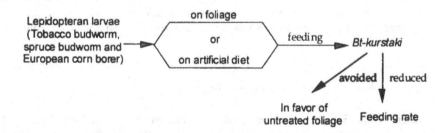

Figure 4.1 Lepidopteran larvae reduce feeding on foliage or artificial diet treated with *Btk.*

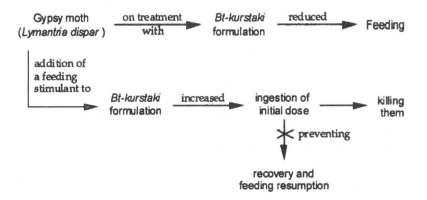

Figure 4.2   Feeding stimulants improve palatability of *Bt* formulation.

Bartlet et al. (1990) investigated the palatability of a starch matrix based granular formulation of *Bt-berliner* by European corn borer larvae and found the larvae seldom fed these. On the other hand, the granule acceptability improved considerably, with addition of a mixture containing lipid, sugar and protein. Coax, a commercially available phago-stimulant (Table 4.3) was most effective in improving acceptability, increasing it 320-fold. Further, when Coax was added to granules, the amount of *Bt* could be reduced by 75%, without reducing mortality, indicating that larvae fed long enough to be killed. A similar response under field conditions was confirmed by McGuire et al. (1990). In a comparative study of the effect of commercially available nutrient based phago-stimulants on various lepidopterous insect larvae, it was observed that Pheast, containing substantial amounts of carbohydrates (39%) and protein (44%), produced best response for all insect species (Farrar and Ridgway, 1994). A list of commercially available phago-stimulants, their sources and major constituents are given in Table 4.3.

3.2.2   *Presence of anti-feedant with* Bt *products*

Basedow and Peters (1997) reported the effect of individual formulations of azadirachtin (Neem Azal®) and *Bt-tenebrionis* (Novodor®) on control of Colorado potato beetle (*Leptinotarsa decelineata*) in potato. They also studied the effect of their combination by application on consecutive days. The results are summarized in Table 4.4.

Table 4.3   Commercially available phago-stimulant spray adjuvants
(Adapted from Farrar and Ridgway, 1994).

Product	Source	Major Constituents (%)	
		Protein	Carbohydrate
Pheast	Agrisense, Fresno, CA	43.8	39.0
Coax	CCT, Litchfield Park, AZ	32.8	50.5
Gusto	Atchem-North America, Philadelphia, PA	34.1	57.7
Consume[a]	Fermone, Phoenix, AZ		
Entice	Custom Chemicals, Frenso, CA	35.3	54.8
Mo-Bait	Loveland Industries, Greeley, CO	2.8	95.0

(a) Identical product to Gusto

Azadirachtin is an effective antifeedant and insect growth-regulating agent. The results indicate that combination of azadirachtin and *Bt-tenebrionis* as a single spray each reduced their individual effectiveness significantly. It appears that azadirachtin has an antagonistic effect on *Bt-tenebrionis* due to reduction or near stoppage after initial feeding by the larvae. It is apparent that insects distinguish antifeedants as they distinguish feeding stimulants.

Table 4.4   Effect of Azadirachtin and *Bt-tenebrionis* formulations on control
of Colorado potato beetle.

Formulation	Control of Colorado potato beetle (%)	Feeding damage (%)	Yield (potato) (Kg/20 plants)	
Bt-tenebrionis (Novodor 3l/ha)	55.4	20	11.20	(119%)
Azadirachtin (Neem Azal 0.5%)	71	7	12.14	(129%)
Combination (5%) (Neem Azal 0.1%+ Novodor 1.5 l/ha)	5	44	10.57	(112%)
Untreated	0	63	9.45	(100%)

3.3 Non-Toxic and Mildly Toxic Additives: Potentiation of the Activity of Bt-Toxins

3.3.1   Protease inhibitors   The addition of protease inhibitors to protect the proteinaceous toxin is reported to potentiate *Btk* and *Bti* activities many fold. MacIntosh et al. (1990) reported the addition of extremely low levels of protease inhibitors enhanced the insecticidal activity of *Btk* by 2-20 fold against a variety of lepidopterans. A similar effect was seen in *Bti* formulation tested in combination with soybean protease inhibitor against *Aedes aegypti*, potentiating activity 3 fold. It is postulated that protease inhibitors may inhibit the degradation of membrane-bound receptors, thus increasing the half-lives and the ability to bind *Bt* proteins.

3.3.2   Allelo-chemicals   Plants produce allelo-chemicals to resist insect attack. These chemicals could also increase the efficacy of biological control agents. Tannins, a major constituent of *Texus* sp. causes larval mortality. Addition of tannic acid, an expensive commercial source of tannin, was tested to increase the efficacy of sub-lethal concentration of *Bt*. Supplementation with 25-500 ppm tannic acid yielded 55-75% mortalities of *Trichoplusia ni*. In comparision, *Bt* alone produced only a 10% mortality (Gibson et al., 1995).

3.3.3   *Non-toxic chemicals*   Effectiveness of *Bt-berliner* on addition of several non-toxic chemicals has been reported against *Spodoptera* (Salama et al., 1985). Inorganic salts (eg. carbonates of sodium. potassium and calcium, sulphates of magnesium and zinc), boric acid, lipid emulsifying agent such as Tween 60 and protein solublizing agents have been found to enhance efficacy significantly. These findings have been corroborated in field trials. It was found that addition of 0.075% potassium carbonate, zinc sulphate or calcium carbonate to the microbial suspension, lead to a greatly enhanced efficacy of *Btk* formulation (DiPel) against *S. littoralis* on soybean, both in terms of larval population reduction and crop yield increase (Morris et al. (1995).

3.3.4   Effect of addition of EDTA   Lin and Tabashnik (1997) reported interesting observations of EDTA addition to *Btk*. They observed that feeding of EDTA by itself to diamondback moth larvae did not result in mortality, but *Btk* along with EDTA reduced LC_{50} by 2-3 fold in the resistant larvae (both neonates and 3rd instars). In case of susceptible larvae, combination with EDTA did not lead to reduction of LC_{50} of *Btk* towards 3rd instars, although in case of neonates, there was a 5-fold reduction in LC_{50} of *Btk* (Figure 4.3). It is clear that addition of EDTA significantly increased the toxicity of *Btk*.

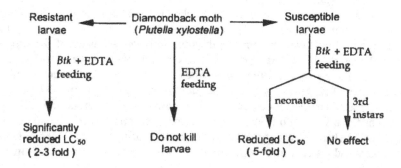

Figure 4.3   Effect of addition of EDTA on the efficacy of *Btk* formulation.

3.4   *Protein 'Enhancin' Synergises* Bt

'Enhancin' is a 104 KDa enzymatic protein which has been found to synergise activity of *Bt* as well as nuclear polyhedrosis viruses (NPVs) in noctuid moths at levels of 50 picograms per insect (Menn, 1996). The interior lining of the insect's intestine, the peritrophic matrix (made mostly of chitin and proteins) is an important component in the insect's immune system against certain microbial attack (biological control agents). Enhancin, when consumed by the insect, binds to a major mucinous protein of peritrophic matrix and destroys the matrix structure. Thus, apparently the passage of biocontrol agents (*Bt*, baculoviruses) is facilitated through this relatively impermeable matrix unimpeded, resulting in the rapid death of the insect.

3.5   *Effect of Adjuvants to Enhance Efficacy*

The ability of sprayed microbial formulation to remain intact on a leaf surface, not withstanding adverse environmental conditions, is a prerequisite to achieving the optimum activity inherent with the microbial system. Factors such as photo-degradation, volatility, spray drift and rain fall, all effectively reduce the potential for the pest control agent to reach the target. *Bt* does not exhibit significant activity beyond 3-4 days on the target foliage, when exposed to sunlight, moisture and leaf exudates. A number of approaches have been tried to protect *Bt* and thereby enhance the persistence on the foliage. Modifications of current *Bt* formulations via the addition of adjuvants have been used as one of the approaches.

3.5.1   Rain fastness enhancement   The addition of commercial wetting, sticking and spreading agents is recommended when sprays are applied to waxy surfaces such as cabbage and broccoli or when rain fastness is important. An oil-based formulation is reportedly less susceptible to wash off by rain, than aqueous-based formulations, presumably due to better adhesion of the droplets (Devisetty et al., 1998). Earlier, these authors had reported the findings of W. Mclane (Cibulsky et al., 1993), that adjuvant 'Rhoplen B60A' improved the rain-fastness of *Bt* on oak seedlings, by demonstrating 43% improved mortality of gypsymoth larvae as compared to *Bt* formulation without the adjuvant. Narayanan and Fanniello (1996) reported improved rain fastness by addition of 'Agrimax 3' (Int'l Speciality Products) composed of pyrrlidone-based solvents, surfactants and water insoluble polymers micro-dispersed in aqueous medium.

3.5.2   Photoprotection enhancement   Sunlight inactivation of *Bt* has been demonstrated to be partly responsible for its short persistence in the field. Pusztai et al. (1987) conducted photo-stability studies with pure δ-endotoxin crystals from *Bt* strains HD-1 and HD-73. Their results demonstrated that the 300-380 nm range of the solar spectrum was largely responsible for loss of toxicity. Cohen et al. (1991) reported photoprotection of the toxic component by adsorption of cationic chromophores, such as acriflavin, methyl green and Rhodamine B to *Bt*. It was reported that acriflavin gave the best protection. A mechanism involving the energy transfer from the excited tryptophan moieties to the chromophore molecules has been suggested. The insect pigment melanin has been found remarkably effective at concentration as low as 0.0003% (Patel et al., 1996). Tamez-Guerra et al. (1996) investigated photo-stability of spray dried gelatinized starch-based granule formulations of *Bt*. They found that addition of lactic or citric acid with starch/flour materials provided increased protection against solar radiation. This was attributed to matrix-like granule formation on spray drying, where the active ingredient is distributed throughout the granule (i.e., active ingredient is on the outside as well as within the granule).

3.5.3   Flow Property of Bt formulations   A good flow property of *Bt* formulations is a requirement for good spray-droplet deposition. A viscosity range of 700 m. pas. or lower has been suggested to give better atomization efficiency (Devisetty et al., 1998). Use of thickening agents can also control drift and evaporation of aerial and ground sprays of *Bt*.

Ignoffo et al. (1976) used a commercial adjuvant to reduce evaporation, increase stability in sunlight and increase larval feeding on foliage treated with formulation. Thus depending on specific requirement, a wetting agent, oil, polymerizing agent, spreader, penetrant, sunlight protectant, humactant and thickner can be selected as adjuvants to increase coverage and effectiveness.

4   Commonly Used Formulations of *Bt*

Bt fermentation slurry concentrate is either dried to technical powder concentrate or formulated as stabilized aqueous suspension. *Bt* spray dried technical powders are processed for particle size control and utilized in both liquid (aqueous and non-aqueous suspensions) and wettable powder (WP), water dispersible granules (WG) and impregnated coated granular formulations. Wettable powders and water dispersible granular formulations are predominantly used for pest control on vegetable, agronomic and fruit crops. Non-aqueous and aqueous suspension formulations are widely used for forestry and raw crops such as cotton and soybeans (Devisetty et al., 1998).

Specific applications also justify surface bonded granular formulations, for example, for penetration of dense foliage, for mosquito larvae control or as briquettes for hand application in an aquatic environment where spray application is not practical. The relative merits of various formulation types used for *Bt*-based pest control products are given as follows:

(a)   Wettable powder   Wettable powders are dry, free flowing powders, containing a high concentration of active ingredient(s). They are formulated to facilitate mixing with water into a final spray. The quality of water dispersible powder is judged by the rapidity of wetting when mixed with water, and it's suspensibility in water when mixed in practical dilutions for field application. Fine particle size improves suspensibility. Size reduction is achieved via air milling or hammer milling of the product, to achieve a particle size range of 10-20 µm. Typical examples include DiPel WP (Abbott), Agree (Thermo Trilogy) and Cutlass WP (Ecogen).

(b)   Suspension concentrates (Flowables)   These are preformed suspensions of well-dispersed micronized active ingredients either in water or in oil phase. These are processed by wet milling (Dynomill) of microbial agents along with adjuvants to achieve the particles in 1-2 µm sizes. These formulations have the advantage of being relatively simple to recover and formulate. However, such formulations may also show limited storage stability, perhaps due to higher rate of proton transfer across the microbial cellular membrane in an aqueous medium. Some of the typical examples are Florbac FC (Novo Nordisk), Mattch and MVP II (Mycogen) and VectoBac FC (Abbott).

(c)   Water dispersible granules (WG)   These are formed from all technical powders, i.e. agglomerated with binders. The preferred process of formulating is by fluidized bed granulation of fine slurry. The formulation must wet instantly when put in water and possess good dispersibility characteristics. Commercial WG formulations are stable products with excellent handling and mixing properties. Some of the typical examples include Javelin WG (Thermo Trilogy) and XenTari WG (Abbott). Ecogen has marketed its genetically modified *Bt*-products Crymax and Lepinox also as WG formulations. However Crymax WDG is prepared by extrusion of the cell paste, as against spray drying, giving it desirable handling properties (Baum et al., 1999).

(d)   Granulars   Granular formulations of a microbial system frequently use an inert carrier with a specific size. The microbial agent is bonded to the carrier with various types of sticking agents. These formulations are useful for aquatic larvae control to treat larval habitats under vegetative cover. Examples include DiPel 10G (Abbott) and M-Peril (Mycogen).

(e)   Briquettes   A briquette is a solid block formulation with a diameter ranging from 2 to 6 cms. Briquettes are generally formed by mixing the active ingredients with light density inert granules and binding agents. These formulations allow timely release of individual granules and are useful for the spreading of microbials into water.   Briquettes are convenient for hand applications into aquatic environments where spray applications are not available or not functional. A typical example is Bactimos briquette (Summit Chemical, Baltimore).

5   Improved *Bt* Formulations

With advances in culture, production, formulation and application technologies, *Bt* formulations have undergone significant improvements from the conventional low-potency formulations. Production of high-potency fermented technical materials has contributed in good measure to this development. The development of improved formulations also provided a means to increase the effectiveness of entomopathogens. Pioneering work has been done in forestry for the control of gypsy moth (*Lymantria dispar*), spruce budworm (*Choristoneura fumiferana*), and other forest caterpillars. Optimized droplet size and distribution in a forest canopy from various aircraft, combined with undiluted, high potency formulations have resulted in greatly improved control of lepidopterous forest pests. Particularly, oil-based emulsifiable suspensions and encapsulation techniques have helped the pest control agent reach the target and remain there by withstanding the environmental occurrences.

5.1   Oil-Based Emulifiable Suspensions (ES)

The development of ULV *Bt* formulation of high potency as stable oil emulsifiable suspension (ES) formulation has provided more consistent insect control than dry wettable powder formulations or low potency aqueous-based formulations. The oil-based formulation can be applied directly, without dilution with water and provides superior spray coverage on target foliage. The formulation potency ranging from 8.45 Billion International Units (BIU)/l to 32.82 BIU/l also results in improved efficacy due to presence of concentrated dose of *Bt* in each spray droplet. These formulations are compatible in both water and oil systems and can be tank mixed with many oil based chemical pesticide formulations and non volatile diluents such as cotton seed oil. It has been reported that expansion of *Bt* from original vegetable crop applications to field crops such as cotton and soybean and forest insect control programs in hard wood and coniferous has been facilitated with these developments (Cibulsky et al., 1993). These formulations are also reported to maintain excellent biostability during storage (Devisetty et al., 1998). Some of the examples of emulsifiable oil suspension concentrates include DiPel ES (Abbott), Condor XL (Ecogen) and Delfin ULV (Thermo Trilogy).

The *Bt* usage has increased in combination with non-evaporating spray oils and other additives in cotton, corn and soybean insect control. Thus, *Bt* provided effective control of cotton insects such as tobacco budworm (*Heliothis virescens*) and the bollworm (*Helicoverpa zea*). (Figure 4.4). Similarly, aerial

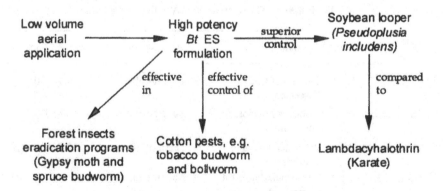

Figure 4.4   High-potency low volume application of *Bt* provides effective control of insects in agriculture and forestry.

application of *Bt* ES formulation for control of soybean insect control, specifically for soybean looper (*Pseudoplusia includens*) have produced equivalent or superior control to synthetic pyrethroid lambdacyhalothrin (Karate® – Zeneca)

Some of the other improved *Bt* formulations that are available include high-potency aqueous flowable concentrates, non-aqueous emulsifiable suspension formulations and non-emulsifiable oil suspension formulations. These are highly suitable for low volume (4.7 to 18.7 l/ha) and ultra-low volume (< 4.7 l/ha) applications on agronomic crops and for forestry use. Some commercially available formulations are listed in Table 4.5.

5.2   Encapsulated Formulations

5.2.1   Starch matrix based encapsulation: Dunkal and Shasha (1988) reported encapsulation of *Bt* in a granular starch matrix that improved the residual activity of *B. thuringiensis* for the European corn borer (*Ostrinia nubilalis*). Shasha and Dunkel (1989) patented a process for encapsulating biocontrol agents such as pathogenic bacteria and viruses in a protective starch matrix without the use of chemical cross linking agents. The biocontrol agent is blended into a dispersion of pre-gelatinized starch, which is then subjected to conditions suitable for retrogradation. The dispersions can be formulated either for recovery of dry granules or as sprayable liquids. McGuire and Shasha (1995) also reported a sprayable starch encapsulation incorporating starch and sucrose into a water dispersible formulation by employing a different process. This starch-based formulation was found effective in maintaining residual activity of *Bt* on cabbage.

Table 4.5   Commercial *Btk* LV /ULV formulations

Trade name	Formulation	Manufacturer	Potency
DiPel ES	Emulsifiable oil suspension	Abbott Labs.	16.91 BIU/L
Foray 48 B	Aqueous flowable concentrate	Novo Nordisk	12.86 BIU/L
Thuricide 48 LV	Aqueous flowable concentrate	Thermo Trilogy	12.70 BIU/L
DiPel 12L	Non-emulsifiable oil suspension concentrate	Abbott Labs.	

(a)   BIU = Billion International Units

5.2.2   CellCap® encapsulation system: Mycogen Corporation (San Diego, California) produced encapsulated formulations of increased residual, high potency bioinsecticides based on their CellCap® encapsulation system. In this process, the single *Bt*-toxins are introduced in a microbial host *Pseudomonas fluorescens*. The recombinant *P. fluorescens* cells are subsequently killed by a proprietary chemical treatment that cross-links the bacterial cell wall to yield a non-viable encapsulated bacterium surrounding the crystal protein (see figure 3.3). An example is MVP®II, a product based on Cry1Ac δ-endotoxin of *Bt-kurstaki*. In a comparative evaluation, Mycogen (1998) reported that MVP II® provided more persistent, long lasting control of caterpillar pests than conventional *Bt* products such as DiPel® and Javelin® (Figure 4.5). The exact mechanism by which the cell protects the biotoxin is unknown, but the fixed cell wall provides a mechanical barrier and may have the ability to impede a variety of denaturing effects, including inactivation by sunlight (Gaertner et al., 1993).

5.2.3   Matricap™ encapsulation: Levy et al. (1996) reported a 'Matricap' granular formulation that encapsulated biolarvicides such as *Bt-israelensis*, for slow release through coating-regulated controlled delivery system from floating or submerged matrices. These formulations have been found useful in targeting the feeding zones of larvae of *Anopheles*, *Aedes* and *Culex* mosquitoes in fresh and brackish water.

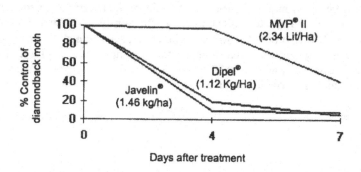

Figure 4.5   Foliar persistence of MVP® II formulated *Bt* versus conventional formulated *Bt*s (Adapted from Mycogen, 1998)

6   Target-Specific Tailor-Made Formulations of Bacterial Larvicides

Anopheles mosquitoes commonly breed in stagnant rain water pools, open storage tanks and other water holding containers. The larvae are adapted to collect particulates from air-water interface. An effective microbial formulation could make use of the feeding behaviour of *Anophelines* by suspending the toxin only in the feeding horizon, i.e. at or near the water surface. Other desirable features of the formulation could include an ability to penetrate barriers, such as natural films, vegetation etc. and remain floating for a long duration without degradation.

The standard water diluted sprayable formulations such as wettable powders and suspension concentrates disperse in the whole water body and soon settle to the bottom. Thus, significant quantities of toxin particles will be outside their feeding zone. Treatment with granules, pallets and briquettes are not appropriate in this situation. The settling feature of the formulations render them ineffective against larvae hatching out of eggs laid soon after treatment. Thus, there is a need for specific formulation(s) for this particular situation. It would appear that a floating type of formulation may increase activity by remaining in the feeding zone much longer.

Ramdas and Khetan (1990) reported a tailor-made self-spreading oil formulation of microbial insecticides that displayed many of the desired properties. The product, a stabilized suspension of the micronized toxicant in an oil phase containing lyophilic surfactants dispersed in an alcohol was found suitable for ultra low volume (ULV) application. The droplets of the product on contact with water spread spontaneously with a great force (spreading pressure >60 dynes/cm^2) into a micro-reticulum (figure 4.6a) that soon began expanding (figure 4.6b), ultimately breaking down to evenly.distributed discrete micro-globules (figure 4.6c). The micro-globules were of optimal dimensions (~5 μm) for rapid tapping and ingestion by the mosquito larvae (figure 4.6d). The formulation was found to be effective at a fractional concentration as compared to an aqueous flowable formulation.

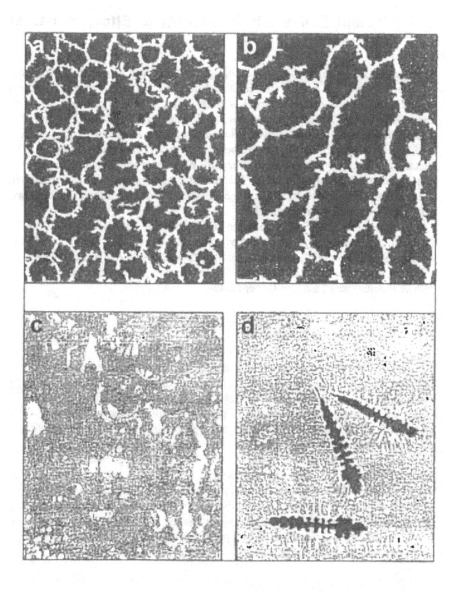

Figure 4.6   Tailor-made self-spreading oil formulation of *Bti* keeps the toxin
particles floating in the larval feeding zone

7   Efficient Delivery is Equivalent to Effectiveness at Low Dose

Application technology, timing, rate of application, drop size, density (number of droplets per cm^2) and weather conditions can effect control of insect population. Method of application has been found to have significant effect on the efficacy of *Bt*. Observations on field application of *Bt* products for control of diamondback moth (*Plutella xylostella*) employing a knapsack, a drop nozzle and an electrostatic sprayer have been reported (Perez et al., 1995). The electrostatic technique showed significantly better performance as compared to other two application techniques. Electrostatic application possibly results in greater deposition of electrically charged droplets on the plant than that obtained with conventional hydraulic nozzles.

Earlier, Law (1983) reported application of *Btk* to control *T. ni* in broccoli. Application through an electrostatic sprayer required only 1/2 - 1/3 the recommended dose per hectare compared with a conventional sprayer. The high efficacy obtained with the electrostatic application device at low rates will reduce the number of treatments per season.

8   Concluding Remarks

Microbial formulations are at the very end of the downstream processing part of the production of microbial larvicides. Effective microbial formulation depends upon a variety of complex areas including biology and feeding behavior of insect pests, ecological requisites of species, individual control agents, their effectiveness and their physico-chemical properties. Rigorous demands are placed on features such as effective dose delivery, palatability, particle size, placement and persistence, stability over a sustained period, and cost. A judicious use of formulation technology, e.g., oil suspension emulsions in combination with application technology, such as low volume application or electrostatic application can be effective in enhancing activity and prolonging the useful life of microbials as pest control agents. Finally, for the optimum use of the specificity of microbials tailor-made formulations are the key rather than multi-purpose standard formulations.

References

1.   Bartlet, R. S., M. R. McGuire and D. A. Black , 1990. "Feeding stimulants for the European corn borer (Lepidoptera: Pyralidae): Additives to a starch based formulation for *Bt*", *Environ. Entomol.*, 19,182-189.

2.   Burges, H. D., and K. A. Jones, 1998. "Formulation of bacteria, viruses and protozoa to control insects", in *Formulation of Microbial Pesticides: Beneficial Microorganisms, Nematodes and Seed Treatments*, H. D. Burges, ed., Kluwer, Dordrecht, The Netherlands, pp. 33-127.

3.   Basedow, T. and A. Peters, 1997. "Control of Colorado potato beetle (*Leptinotarsa decemlineata* say) by an azadirachtin formulation (NeemAzal T/S) by *Bt-tenebrionis* (Novodor) and by combinations of both: short term and long term effects, *Pract. Oriented Results Use Prod. Neem-Ingredients Pheromones, Proc. Workshop*,5th, 1996, H. Kleeburg and G. Zebitz, eds, P.W. Druck and Graphic, Giessen, Germany, pp. 59-65.

4.   Baum, J. A., T. B. Johnson and B. C. Carlton, 1999. *"Bacillus thuringiensis*: Natural and Recombinant bioinsecticide products", *Biopesticides: Use and Delivery*, F.R. Hall and J. J. Mann, eds., Humana Press, Totowa, N.J. pp. 189-209.

5.   Burges, H. D. and K. A. Jones, 1998. "Formulation of bacteria, viruses and protozoa to control insects", in *Formulation of Microbial Pesticides: Beneficial microorganisms, nematodes and seed treatments*, H. D. Burges, ed., Kluwer, Dordrecht, The Netherlands, pp. 33-127.

6.   Cibulsky, R. J., B. N. Devisetty, G. L. Melchiord and B. E. Melin, 1993. "Formulation and application technologies for microbial pesticides: review of progress and future trends", *Jour. Testing & Eval.* 21,500-503.

7.   Cohen, E., H. Rozen, T. Joseph, S. Braun and L. Margulies, 1991. "Photoprotection of *Bacillus thuringiensis kurstaki* from ultraviolet radiation", *J. Invertebr. Pathol.*, 57, 343-351.

8.   Devisetty, B. N., Y. Wang, P. Sudershan, B. L.. Kirpatrick, R. J. Cibulsky and D. Birkhold, 1998. "Formulation and delivery systems for enhanced and extended activity of biopesticides", *Pesticide Formulations and Application Systems: Eighteenth Volume*, J. D. Nalewaja, G. R. Goss and R. S. Tann, eds., American Society for Testing and Materials, Philadelphia, pp. 242.

9.   Dunkal, R. L. and B. S. Shasha, 1988. "Starch encapsulated *Bacillus thuringiensis*: a potential new method for increasing environmental stability of entomopathogens", *Environ. Entomol.*, 17, 120-126.

10.   Farrar, R. R., Jr. and R. L. Ridgway, 1995. "Enhancement of activity of *Bt. berliner* against four lepidopterous insect pests by nutrient based phagostimulants", *J. Entomol. Sci.*, 30, 29-42.

11.   Farrar, R. R., Jr., and R. L. Ridgway, 1994. "Comparative studies of the effects of nutrient-based phago-stimulants on six lepidopterous insect pests", *J. Econ. Entomol.*, 87, 44-52.

12.   Gibson, D. M., L. G. Gallo, S. B. Krasnoff and R. E. B. Ketchum, 1995. "Increased efficacy of *Btk* in combination with tannic acid", *J. Econ. Entomol.*, 88, 270-277.

13.   Gaertner, F. H., T. C. Quick and M. A. Thompson, 1993. "CellCap®: An encapsulation system for insecticidal biotoxin proteins", in *Advanced Engineered Pesticides*, Leo Kim, ed., Marcel Dekker, Inc., New York, pp. 73-83.

14.   Gould, F., A. Anderson, D. Landis and H. Van Mallaert, 1991. "Feeding behavior and growth of *Heliothis virescens* larvae (Lepidoptera: Noctuidae) on

diets containing *Bacillus thuringiensis* formulations or endotoxins", *Entomol. Exp. Appl.*, 58, 199-210.

15.   Khetan, S. K., M. A. Ansari, P. K. Mittal and A. K. Gupta, 1996. "Laboratory and field bioefficacy studies on *Bacillus thuringiensis* var. *israelensis* (Serotype H-14) against *Culex* and *Anopheline* larvae", Unpublished Results.

16.   Law, S. E., 1983. "Electrostatic pesticide spraying: concepts, and practice, *IEEE* (Inst. Electr. Electron. Eng.)*Trans. Ind. Appl.*, IA-19, No. 2.

17.   Levy, R., M. Nicols and W. R. Opp, 1996. "New Matricap pesticide delivery systems", *Proc. Int'l Symp. Control Rel. Bioact. Mater.*, Kyoto, Japan, Control Rel. Soc., 23.

18.   Lin, Y-B and B. E. Tabashnik, 1997. "Synergism of *Bacillus thuringiensis* by EDTA in susceptible and resistant larvae of diamondback moth (*Lepidoptera : Plutellidae*)",*J. Econ. Entomol.*, 90, 287-292.

19.   MacIntosh, S. C., G. M. Kishore, F. D. Perlak, P. G. Marrone, T. B. Stone, S. R. Sims and R. L. Fuchs, 1990. "Potentiation of *Bt* insecticidal activity by serine protease inhibitors", *J. Agric. Food Chem.*, 38, 1145-52.

20.   McGuire, M. R., B. S. Shasha, C. E. Eastman and H. Oloumi-Sadeghi, 1996. "Effect of starch and flour based sprayable formulations on rainfastness and solar stability of *Bacillus thuringiensis*", *J. Econ. Entomol.*, 89, 863-869.

21.   McGuire, M. R. and B. S. Shasha, 1995. "Starch encapsulation of microbial pesticides", in Biorational Pest Control agents: Formulation and Delivery, F. R. Hall and W. J. Barry, eds., American Chemical Society, washington, DC, pp.229-237.

22.   McGuire, M. R., B. S. Shasha, L. C. Lewis, R. J. Bartlet and K. Kinney, 1990. *J. Econ. Entomol.*, 83, 2207-2210.

23.   Menn, J. J., 1996. "Biopesticides: Has their time come?", *Environ. Sci. Health*, B31(3), 383-389.

24.   Molloy, D., S. P. Wright, B. Kaplan, J. Gerardi and P. Peterson, 1994. "Laboratory evaluation of commercial formulations of *Bacillus thuringiensis* var. *israelensis* against mosquito and black fly larvae", *J. Agric. Entomol.* 1, 161-168.

25.   Morris, O. N., V. Converse and P. Kanagaratnam, 1995. "Chemical additive effects of the efficacy of *Bt-berliner subsp. kurstaki* against *Memestra configurata* (*Lepidoptera: Noctuidae*)", *J. Econ. Entomol.*, 88, 815-824.

26.   Mycogen Tipsheet - MVP®II, 1998. http://www.mycogen.com/graphic/ pest /tipsheet /mvp/2213.htm.

27.   Narayanan, K. S. and R. M. Fanniello, 1996. "Superior multipurpose adjuvant system for rainfastness and UV protection", in *Pesticide Formulation and Application Systems*, Vol. 15, H. M. Collins, F. R. Hall and M. Hopkinson, eds., American Society for Testing of Materials.

28.   Patel, K. R., J. A. Wyman, K. A. Patel and B. J. Burden, 1996. "A mutant of *Bacillus thuringiensis* producing a dark brown pigment with increased UV resistance and insecticidal activity", *J. Invertebr. Pathol.*, 67, 120-124.

29.   Perez, C. J., A. M. shelton and R. C. Derksen, 1995. "Effect of application technology and *Bt* subspecies on management of *Bt-kurstaki* resistant diamondback moth (*Lepidoptera : Plutellidae*)", *J. Econ. Entomol.*, 88, 1113-1119.

30. Pusztai, M., P. Fast, H. Kaplan and P. R. Carey, 1987. "The effect of sunlight on the protein crystals from *Bacillus thuringiensis* var. *kurstaki* HD1 and NRD12: A Raman spectroscopic study", *J. Inveribr. Pathol.*, 50, 247-253.

31. Ramdas, P. K. and S. K. Khetan, 1990. "A self-spreading formulation for the control of surface feeding/ inhabiting aquatic pests", *7th International Congress of Pesticide Chemistry*, Abstracts, H. Frehse, E. Kesseler-Schmitz and S. Conway, eds., 2, 39.

32. Rashed, S. S. and M. S. Mulla, 1989. "Factors influencing ingestion of particulate materials by mosquito larvae (*Diptera: Culicidae*)", *J. Med. Entomol.*, 26, 210-216.

33. Salama, H. S., M. S. Foda and A. Sharaby, 1985. "Potential of some chemicals to increase the effectiveness of *Bt-berliner* against *Spodoptera littoralis*", *Z. Aug. Ent.*, 100, 425-433.

34. Shasha, B. S. and R. L. Dunkle, 1989. "Starch encapsulation of entomopathogens", U. S. Patent 4,859,377.

35. Shieh, T. R., 1998. "The formulation of biologicals", in *Pesticide Formulation: Recent Developments and Their Application in Developing Countries*, J. W. Van Valkenburg, B. Sugavanam and S. K. Khetan, eds., New Age International, New Delhi. pp. 232-248

36. Skovmand, O., D. Hoegh, H. S. Pederson and T. Rasmussen, 1997. "Parameters influencing potency of *Bacillus thuringiensis* var. *israelensis* products", *J. Econ. Entomol.*, 90, 361-369.

37. Tamez-Guerra, P., M. R. McGuire, H. Medrano-Roldan, L. J. Galan-Wong, B. S. Shasha and F. E. Vega, 1996. "Sprayable granule formulations for *Bacillus thuringiensis*", *J. Econ. Entomol.*, 89, 1424-1430.

5

Insect Resistance to *Bt* Toxins

1   Introduction

Resistance is a genetically based decrease in susceptibility of a population to an insecticide. As resistance entails changes in allele frequencies in a population, by definition, it is an evolutionary phenomenon. Insects are known for their ability to develop resistance to certain insecticides rapidly. Resistance occurs particularly when insecticides are used repeatedly and at high concentrations leading to intensive selection pressure on insect populations. More than 500 species of insects and mites are known to have developed resistance to insecticides and miticides (Georghiou and Lagunes-Tejeda, 1991).

To document development of resistance, one must show that repeated treatments with an insecticide have caused a significant increase in the amount of insecticide required to kill a certain proportion of population (e.g. the concentration that kills 50%, LC_{50}) or a significant decrease in the percentage mortality caused by a fixed amount of insecticide. The degree of resistance observed in an insect population is typically expressed as the resistance ratio, i.e. number of LC_{50}-resistant insects / number of LC_{50}-sensitive insects.

Laboratory colonies of more than 15 different insect pests have been demonstrated to develop resistance to *Bt* proteins, including Indian meal moth, tobacco budworm, beet armyworm, pink bollworm and Colorado potato beetle. Moreover, the diamondback moth, a world-wide pest of cole crops, has developed high levels of resistance to *Bt* insecticide in field populations in several countries.

Insects could possibly achieve resistance to *Bt* toxins by different mechanisms ranging from the point of protoxin ingestion to the insertion of toxin in the membrane. The factors affecting the binding of toxin to the receptor would result in selective resistance to one particular *Bt* toxin. On the other hand, those steps utilised by all the toxins viz. proteolysis of protoxins, conformational alterations and membrane insertion may lead to cross-resistance to multiple *Bt* toxins. Reduced binding of *Bt* toxin to the brush border membrane of the midgut epithelium was identified as a primary mechanism of resistance to Indian meal moth and diamondback moth.

Several authors have reviewed different aspects of insect resistance to *Bt* toxins and its management (Schnepf et al., 1998; Bauer, 1995; Tabashnik, 1994; Marrone and Macintosh, 1993; and McGaughey and Whalton, 1992). In this chapter, an overview of resistance to *Bt* toxins has been presented. The discussion includes the synergistic interactions between various Cry proteins and Cry proteins and spores. Current views and approaches on resistance management have also been covered here.

2   Resistance to *Bt* toxins in *Lepidoptera*

More than two decades of frequent applications of *Bt*-based formulations in many parts of the world, have resulted in the development to *Bt* spore-crystal toxins complex resistant populations of lepidopteran caterpillars. It is known that toxicity of *Bt*-toxins relates to their ability to bind to receptors in the larval mid-gut and a single insect may have different receptors for different *Bt*-toxins. There has been evolution of insect resistance to insecticidal proteins and it has been found that resistance is toxin-specific. The mechanism of resistance relates to change in binding characteristics of *Bt*-ICP against Lepidoptera, due to either a reduction in the concentration of microvillar membrane receptors, reduced toxin affinity for the receptors, or an implied failure of the toxin to insert into the microvillar membrane after binding. A decrease in the affinity between the toxin and target receptors is frequently associated with a mutation on the target receptor (Figure 5.1).

McGaughey (1985) observed that Indian meal moth (*Plodia interpunctella*) populations from grain storage bins that had been treated for 1 to 5 months with a *Bt-kurstaki* formulation had a small but significant increase in LC_{50}s relative to populations in untreated bins. Further, he demonstrated in laboratory experiments, that after 15 generations of selection, insects from the treated colony showed LC_{50}s nearly 100-fold greater than those shown by control experiments. Van Rie et al. (1990) demonstrated that the resistance

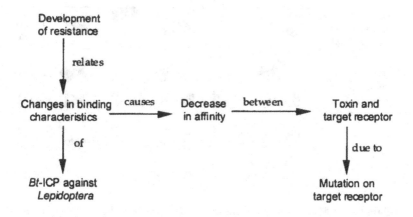

Figure 5.1   Mechanism of resistance development in *Lepidoptera* against
Bt through decrease in affinity.

correlated with a significant reduction in affinity of the membrane receptor for
the Cry1Ab protein, one of the toxins present in the *Bt-kurstaki* formulation
(DiPel®) used for selection (Figure 5.2). Evidence was also found that in
resistant *P. interpunctella* larvae, in addition to changes in binding, there may be
decrease in cleavage of the full-length *Bt* protein to its toxic form due to a loss
of protease function (Oppert et al., 1997).

The Indian meal moth (*Plodia interpunctella*), as discussed, has
probably evolved low levels of resistance in stored grain treated with *Bt*
(McGaughey, 1985). Resistance in another stored grain pest almond moth
(*Cadra cautella*) has been reported only from laboratory studies (McGaughey
and Beeman, 1989). Similarly, resistance development in tobacco budworm
(*Heliothis virescens*) (Stone et al., 1989, Gould et al., 1992 and Lee et al., 1995)
and the Colorado potato beetle (*Leptinotarsa decerulineata*) (Whalon et al.,
1993) has been also reported from laboratory selection experiments. The
resistance trait proved to be recessive. When selection was removed before
resistance became fixed, resistance levels decreased. However, the very
possibility of resistance development in the field populations is of great concern
because of their economic significance.

Although, at least eight species have adapted to resist *Bt* δ-endotoxins,
only one species, diamondback moth (*Plutella xylostella*) has actually developed
significant levels of resistance under field conditions. The mechanism of
resistance in diamondback moth population that had developed resistance in the
field was demonstrated to be also due to change in ICP membrane receptor
(Ferre et al. 1991). It has been found that resistant populations of diamondback

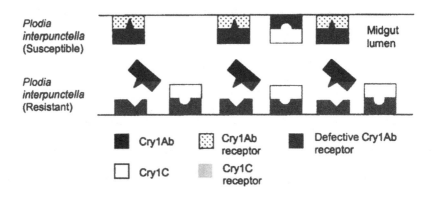

Figure 5.2   Mechanism of resistance to ICPs in the Indian meal moth (*Plodia interpunctella*). The brush border membrane of the midgut of wild-type (susceptible) larvae carries receptors for Cry1Ab and Cry1C toxins that make them sensitive to these toxins. Resistance to Cry1Ab in *Plodia* (resistant) is due to defective receptor for the toxin. The Cry1Ab resistant larvae are still sensitive to Cry1C, which binds to another receptor. No cross-resistance is observed (Lambert and Peferoen, 1992).

moth to Cry1A toxins from *Bt-kurstaki*, showed cross resistance to Cry1F, but not to Cry1B or Cry1C toxins (Tabashnik et al. 1996). It has also been reported that diamondback moth with populations treated with *Bt-aizawai* displayed resistance to Cry1C (Liu et al., 1996). On HPLC analysis of commercial formulations of *Bt-kurstaki* and *Bt-aizawai*, it was apparent that Cry1Ab was the most abundant Cry1 protein in both, while Cry1C was present in *Bt*-aizawai alone. The Cry1C resistant diamondback moth populations were found resistant to Cry1Ab, but this was lower than in a Cry1C susceptible colony that had been selected by *Bt-kurstaki* (Figure 5.3). The toxins in the cross-resistance group have significant amino acid sequence similarity in domain II, a region believed to be important for receptor binding in many systems (see Chapter 1).

2.1   Bt *Spore-Crystal Interactions*

For many lepidopteran insect pests, such as the beet armyworm (*Spodoptera exigua*), the Bt spores present on bioinsecticide formulation also contribute substantially to toxicity (Moar et al., 1995). The synergistic effect of spores has also been reported for gypsy moth *(Lymantria dispa)* (DuBois and Dean, 1995).

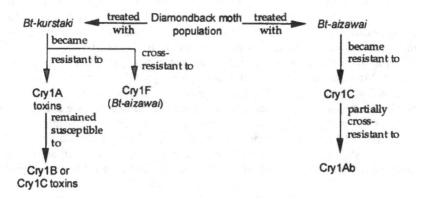

Figure 5.3   Resistance to *Bt*-toxins in diamondback moth.

A study of laboratory colony of *Spodoptera littoralis* treated with *Bt*-spore crystal preparations of Cry1C toxin, developed a resistant population (>500 fold resistance) in 14 generations. This population exhibited partial cross-resistance to Cry1D, Cry1E, Cry1Ab toxins and to parental strain *Bt-aizawai*. However, their susceptibility to Cry1F was unchanged (Muller-Cohn et al., 1996) (Figure 5.4).

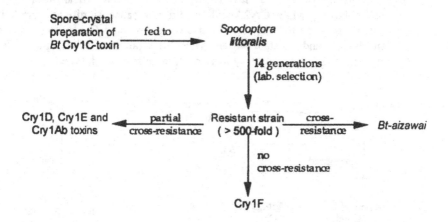

Figure 5.4   Cry 1C-resistant *S. littoralis* shows partial cross-resistance to other Cry-toxins.

Tang et al. (1996) studied toxicity of *Bt* spore and single crystal protein, to resistant population of diamondback moth. The diamondback moth larvae tested, showed a high level of field-evolved resistance to *Btk* spore and all the three Cry1A protoxins, but still susceptible to Cry1B, Cry1C and Cry1D. When *Bt-kurstaki* spore and single cystal protoxin preparation (1:1) was fed to susceptible larval of diamondback moth, spores synergised the activity of Cry1A (5 to 8-fold) and Cry1C (1.7 fold), respectively. On the other hand, when this combination was fed to resistant larvae, spores failed to synergise activity of all three Cry1A protoxins, but synergised the activity of Cry1C (5.3- fold), (Figure 5.5). In a bid to determine possible mechanisms of resistance, binding studies with Cry1Ab, Cry1B and Cry1C were performed. The resistant larvae showed a dramatically reduced binding of Cry1Ab as compared with that in susceptible larvae, but no difference in binding of Cry1B or Cry1C.

Liu et al. (1998) extended the study of interactions between spores and single cystal *Bt*-toxins to interactions of spores and combination of toxins that occur in crystals from *Bt-kurstaki* and *Bt-aizawai*. In leaf residue bio-assays on diamondback moth larvae, they reported synergistic interactions on addition of *Bt-kurstaki* spores to *Bt-kurstaki* crystals. A reduction in LC_{50} of *Bt-kurstaki* crystals against susceptible larvae (10-fold) and resistant larvae (45-fold) respectively, was observed. On the other hand, spores of *Bt-aizawai* had little or no effect on toxicity of crystals of *Bt-aizawai*.

The interactions between spore and single toxins revealed synergism between *Bt-kurstaki* spores and Cry2Aa against susceptible diamondback moth larvae. Preparations of *Bt-kurstaki* spores alone caused significant mortality of susceptible larvae, while Cry2Aa alone did not cause significant mortality. Synergism also occurred between *Bt-aizawai* spores and Cry1C toxin against both susceptible and resistant larvae. Although, again *Bt-aizawai* spores alone did not cause significant mortality in susceptible or resistant larvae.

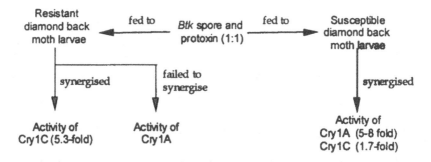

Figure 5.5   Synergistic activity of Cry proteins.

This spore effect on the insecticidal activity of *Bt* may be due to septicemia, the ability of the spore to germinate within the insect midgut to penetrate the disrupted midgut epithelium, and to enter and proliferate within the hoemcoel. For many lepidopteran insect pests, it is therefore desirable that *Bt* bioinsecticide formulations contain a mixture of spores and crystals to achieve maximum efficacy. An implication of this study on *Bt*-toxin expressing transgenic plants and transgenic bacteria is that absence of spores may accelerate evolution of pest resistance. However, such a possibility would need to be experimentally evaluated.

2.2   Bt Cry-Cry Interactions

In a study, the toxicity of toxins CrylAa, CrylAb and CrylAc was found against gypsy moth (*Lymantria dispar*) by force-feeding bio-assay (Lee et al.,1996). CrylAa exhibited higher toxicity than toxins CrylAb and CrylAc. When bioassay was conducted with mixtures of CrylAa and CrylAb, CrylAb and CrylAc, and CrylAa and CrylAc, a synergistic activity was observed with the mixture of CrylAa and CrylAc. On the other hand, a mixture of CrylAa and CrylAb exhibited an antagonistic effect (Figure 5.6).

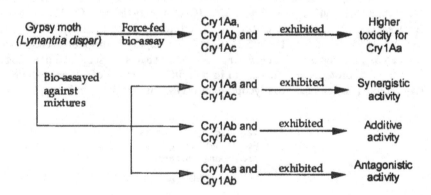

Figure 5.6   Force-feed bio-assay of gypsy moth with individual and mixtures of CrylA proteins.

3   Synergistic Interactions of *Bt-israelensis* Toxins and Effect on Resistance Development

Despite several years of field usage, no significant resistance has been detected in field populations of mosquitoes towards *Bt-israelensis* even in areas where they have been treated intensively. Attempts at laboratory selection have produced only low levels of resistance in *Culex quinquefasciatus* leading to the hypothesis that the heterogeneous mixtures of toxins present in *Bt-israelensis* constitute an effective defence against the development of resistance. *Bt-israelensis* crystals contain four major δ-endotoxins, namely Cry4A, Cry4B, Cry11A (earlier known as CryIVD) and Cyt1A. Crickmore et al. (1995) determined their relative activities through bio-assay against *Aedes aegypti* larvae. Bio-assay of mixtures of the individual toxins revealed a number of synergistic interactions. Poncet et al. (1995) evaluated synergistic interactions among the Cry4A, Cry4B and Cry11A toxic components of *Bt-israelensis* crystals against *Aedes. aegypti, Anopheles stephensi* and *Culex pipiens*. It was found that Cry4A and both Cry4B and Cry11A gave synergistic effect while Cry4B and Cry11A showed a simple additive effect (Figure 5.7). Other crystal toxins presumably contribute to full toxicity of wild type *Bti* crystals. This also explains, as to why native crystal is considerably more toxic than any of the individual toxins.

Wirth, Georghiou and Federici (1997) studied the synergistic interactions of Cyt1A with Cry toxins (Cry4A, Cry4B and Cry11A) of *Bt-israelensis*. They observed that Cyt1A, in addition to being toxic, synergises the toxicity of Cry proteins (Table 5.1). Cyt1A is a highly hydrophobic endotoxin with an affinity for *Bt-israelensis* Cry toxins, but shares no sequence homology with the Cry proteins and appears to have a different mode of action. Whereas, Cry proteins initially bind to glycoproteins on the microvillar membrane,

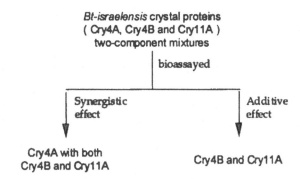

Figure 5.7   Toxic effect of mixture of *Bt-israelensis* Cry proteins.

Table 5.1   Synergism between CytlA and Cry toxins of *Bt-israelensis* tested against larvae of *C. quinquefasciatus* (Wirth et al., 1997)

Toxins	LC_{50}, μg/ml		Synergistic Factor
	Observed	Theoretical	
Cry 11A + Cytl A	0.321	0.978	3.0
Cry4A + Cry4B + Cytl A	0.0501	0.361	7.2
Cry4A + Cry4B + Cry11A + Cytl A	0.0141	0.0380	2.7

primary affinity of CytlA is for the lipid component of the membrane, specifically, for unsaturated fatty acids in the lipid portion of microvillar membrane. It is believed that CytlA may be the key protein accounting for the inability of mosquitoes to quickly develop resistance to *Bt-israelensis*. They also tested the combination of sub-lethal quantities of CytlA with Cry proteins of *Bt-israelensis* and found that it suppressed or markedly reduced high levels of Cry resistance in the mosquito *Culex quinquefasciatus*.

Georghiou and Wirth (1997) also studied the influence of *Bt-israelensis* toxins as a model system to determine the speed and magnitude of evolution of resistance in the mosquito *Culex quinquifasciatus* during selection for 28 consecutive generations with single or multiple toxins. Resistance became evident with single toxin Cry11A, when by 28th generation (F_{28}) resistance ratio (RR) at LC_{95} was >913. Resistance was found to evolve slowly and at low level with two toxins Cry4A and Cry4B, as at F_{25}, RR was found to be >122. It was observed to be lower still with three toxins Cry4A, Cry4B and Cry11A (RR=91 at F_{28}). It was found to be remarkably low with all the four toxins (RR = 3.2) (Table 5.2).

Table 5.2   Development of resistance in *C. quinquefasciatus* with individual and mixture of *Bt-israelensis* toxins (Adapted from Georghiou and Wirth, 1997)

Bt-israelensis toxins	C. quinquifasciatus generation	Resistance ratio at LC_{95}
Cry 11A	F_{28}	>913
Cry 4A + Cry 4B	F_{25}	>122
Cry 4A + Cry 4B + Cry 11A	F_{28}	91
Cry 4A + Cry 4B + Cry 11A + Cytl A	F_{28}	3.2

The study provided evidence for two hypotheses regarding *Bt* resistance: (a) that combinations of *Bt* Cry proteins are better than Cry proteins used alone, and (b) that using a protein like CytA, with a different mode of action, in combination with Cry proteins, may delay the development of resistance.

4   Insect Resistance to *B. sphaericus*

Bacillus sphaericus inclusions are made of equimolar amounts of proteins of 52 and 42 kDa which act as a binary toxin that has a potential for resistance development. It is reported that *Culex quinquefasciatus* develops resistance to *B. sphaericus* at various degrees. Treatments with preparations of *B. sphaericus* at high dosages and high frequency has given rise to resistance development ranging from 30-40 fold to 10,000 fold, depending on the intensity of selection pressure. It has been observed that the resistant strains of larvae of *Culex quinquefasciatus* ingested *B. sphaericus* particulates at a significantly lower rate than did larvae of susceptible strains. Since ingestion of toxin particulate is a necessity for expression of toxicity, it has been suggested that a change in favour of a lower ingestion rate may provide an initial mechanism for emergence of resistance (Rodcharoen and Mulla,1995).

B. sphaericus binary toxin bind to larval midgut cells in *Culex* species with high specificity and high affinity, leading to high susceptibility of these species. The mechanism of resistance to *B. sphaericus* toxin as well as *Bt*-toxins seems to depend on similar non-functionality of membrane midgut receptors (Nielsen-Leroux et al., 1995). However, it has been observed that mosquito populations resistant to *B. sphaericus* remain susceptible to *Bt-israelensis* toxins. In binding experiments, no competition was observed between *B. sphaericus* toxin and *Bt-israelensis* ICP toxins, indicating that different receptors are involved.

5   Resistance Management of *Bt* Toxins

Bt-toxins are used as an environmentally benign foliar spray and are among a handful of microbial biocontrol agents available to conventional growers, who want safe, efficient pest control tools and to organic farmers, who are concerned about loosing one of their most reliable biocontrol tools. The challenge is to develop and implement strategies that provide adequate short-term insect control while delaying insect resistance to *Bt*-toxins. This will ensure that the long-term economic and consumer benefits are preserved and value of this environmentally useful insecticide is not jeopardised.

5.1   Resistance Monitoring

Establishment of baseline susceptibility in field population of target pests toward insecticidal protein, precisely characterised in terms of its nature and amount of δ-endotoxins, would be the first step. Monitoring is also necessary to learn whether a field control failure resulted from resistance or other factors that might inhibit expression of the *Bt* protein. Thus, central to the operation of resistance management program is an effective monitoring program to detect as early as possible, shifts in pest susceptibility that could be abated in the initial stages of resistance problem. Resistance management programs are best implemented before the pest becomes resistant. Detecting resistance may be possible before control failures occur, if monitoring techniques are sensitive enough to provide complete discrimination between resistant and susceptible individuals. A monitoring strategy for early detection of Lepidoptera resistance to *Bt*-ICPs has been reported by employing larval growth inhibition assay using sub-lethal Cry1Ac dose on *Heliothis virescens* and *Helicoverpa zea*. This approach is claimed to be more sensitive than dose-response mortality assays. It allowed visual discrimination of resistant from susceptible phenotypes (Sims et al., 1996).

5.2   General Strategies

Resistance management strategies have been proposed as a means to decrease the rate at which resistance evolves and hence to keep the frequency of resistance genes sufficiently low for insect control. The development of strategy relies heavily on theoretical assumptions and on computer models simulating insect population growth under various conditions. Tabashnik (1994) outlined the strategies in the context of *P. xylostella* – *B. thuringiensis* system. These included, (a) the use of mixtures of toxins with different mechanisms, (b) synergists to increase toxicity, (c) mosaic application to resort to time alternations rather than space alternations, (d) rotation of toxins to reduce the frequency of resistant individuals, (e) ultra-high doses of toxin that kill resistant heterozygotes and homozygotes and (f) refuges to facilitate survival of susceptible individuals.

Perez et al. (1997) tested, some of these strategies that were essentially based on the results of laboratory investigations, on field population of *P. xylostella* with already detectable resistance. A higher dose rate (twice of field rate) of application and use of refuges (equivalent to 25% of area planted) was found ineffective, once resistance was already detected. A refuge is intended to dilute the resistance; however when the refuge is already highly contaminated with resistance alleles, it can not be very effective in dilution. The outcome of rotations of *Bt*-products also did not result on the anticipated lines. A switchover to *Bt-aizawai* was not found to improve the efficacy any significantly in overall reduction of infestations of *P. xylostella* caused by *Bt-kurstaki*. However, continued use of *Bt-kurstaki* would have further increased resistance beyond the existing level. A likely explanation for these results given was that in the early studies *Bt-aizawai* had shown ~ 3-fold cross-resistance with *Bt-kurstaki* resistant *P. xylostella*. Both *Bt* subspecies contain same Cry1A toxins, whereas *Bt-aizawai* also contains Cry1C. Thus minimal cross-resistance may have prevented *Bt-aizawai* from reaching a higher level of efficacy.

5.3 Management of Insect Resistance by Designing Recombinant Bt *Strains*

Generally, resistance to *B. thuringiensis* insecticidal toxins is the consequence of a mutation(s) that alters an insect midgut receptor proteins(s) so that it no longer binds to the Cry protein. However, if a toxin gene was engineered so that toxin bound to other midgut cell surface proteins, then resistance might be less likely to arise (Glick and Pasternak, 1998). For example, the Raven® contains Cry3A and Cry3Bb protein genes, in addition to two *cry1Ac* genes. The two *cry3* genes contribute to a much higher productivity in fermentation of this strain, and have different binding characteristics on Colorado potato beetle midgut cell membranes. In a study conducted jointly by Ecogen scientists and Michigan State University, laboratory-selected potato beetles, that were resistant to one of the Cry3 proteins, showed only minimal resistance to the second Cry3. Thus, in practice, an individual beetle would have to undergo two independent resistance mutations to become resistant to the Raven strain. It was also seen that when the same resistant potato beetle was exposed to a mixture of that Cry3 protein and the Cry1Ac proteins, the Cry 3 resistance is strongly reduced (Figure 5.8). This effect is presumed to be due to Cry-Cry interactions that occurs between the two

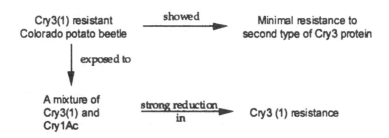

Figure 5.8   Two different strategies for minimising resistance development to Raven *Bt* strain by Colorado potato beetle.

Cry proteins at the level of midgut binding. Thus, the Raven strain is claimed to incorporate two different strategies to minimise the likelihood that the principal insect target would develop resistance to it (Carlton, 1995).

Another approach of resistance management of *Bt* by recombinant technology is by construction of hybrid protein construct. For example, the insecticidal proteins Cry1C and Cry1E are both toxic to Lepidoptera but have different species specificities. Cry1C is active against *S. exigua*, *Mamestra brassicae* and *Manduca sexta*, while Cry1E is active against only *M. sexta*. In an experiment, hybrid Cry1C-Cry1E proteins were constructed and tested for their toxicity to different insect species as well as for their ability to bind to different receptors (Figure 5.9). The hybrid toxin G27, which contained domain III from Cry1C, was toxic to *S. exigua* larvae, even though it bound to the Cry1E receptor but not to the Cry1C receptor. Conversely, the hybrid toxin F26 was not toxic to *S. exigua* larvae even though it bound to the Cry1C receptor. Since the Cry1C and hybrid G27 proteins bind to different insect midgut receptors although both are toxic to *S. exigua*, either simultaneous or alternating treatments of *S. exigua* with these two *Bt* insecticidal toxins might limit the development of strains that are resistant to these toxins. Resistance to both Cry1C and G27 would require mutations in two separate midgut receptor proteins (Bosch et al., 1994).

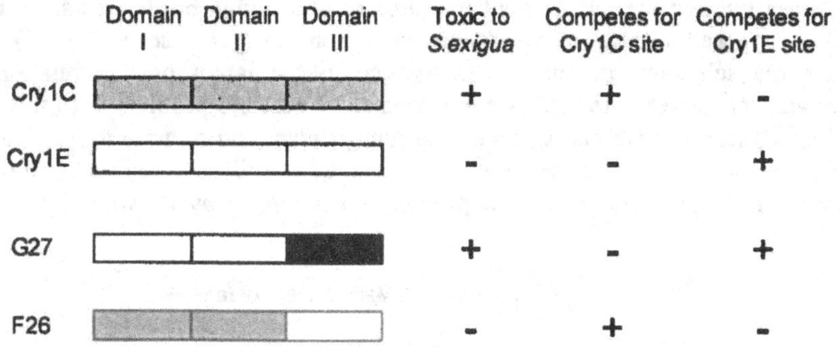

	Domain I	Domain II	Domain III	Toxic to S.exigua	Competes for Cry1C site	Competes for Cry1E site
Cry1C				+	+	-
Cry1E				-	-	+
G27				+	-	+
F26				-	+	-

Figure 5.9   Toxicity and binding specificity of Cry1C, Cry1E, and hybrid toxins G27 and F26 (Adapted from Bosch et al., 1994).

5.4   Management of Insect Resistance in Transgenic Bt-plants

Transgenic crops threaten the viability of *Bt* because, unlike the *Bt* sprays, which last, but a short time, the crops produce toxins throughout their lives. This may result in strong selection pressure with greater probability to resistance development. McGaughey, Gould and Gelernter (1998) summarised the strategies on insect resistance management to *Bt*-transgenic plants. The strategies involved (a) decreasing the proportion of the pest population exposed to a *Bt*- toxin by maintaining areas that are planted to crop cultivars that do not produce any toxin. These areas serve as refuges for insect susceptible to the *Bt*-toxins. These strategies also rely on (b) making sure that pests exposed to the *Bt*-toxin encounter at least 25 times as much toxin as would be needed to kill 99% of susceptible pest species, i.e. a dose expected to kill all or most heterozygotes. In addition, (c) *Bt*-producing plants having two biochemically unrelated toxins, each at a high dose, is also expected to significantly decrease the rate of resistance development (Figure 5.10). Many new strains of *Bt*, or *Bt*-toxins fit the requirements for a multiple toxin strategy. However, many of these toxins are similar enough in activity that resistance to one could cause a degree of cross-resistance to others.

Liu and Tabashink (1997) reported experimental evidence that a refugia strategy, as a source of susceptible insects for mating with the selected populations, to prevent the fixation of resistance, delays insect adoption to *Bt* toxins. Refuges to occur naturally, a certain percentage of an agricultural acreage is planted with non-transgenic plants. Refuges may also be established by design through the use of seed mixtures or susceptible border rows. Menn (1996) proposed mixing of transgenic with non-transgenic seeds (80:20) to reduce the selection pressure for resistance in the target pest populations. However, mixed seed strategy is controversial. A possible problem with mixed seed arises from larvae that survive on a non-*Bt* plant and migrate to *Bt* cotton where they will be less sensitive to *Bt* because of size. This could compromise insect control and increase selection pressure for resistance by allowing the

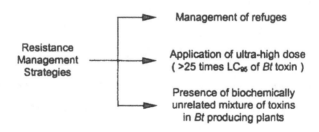

Figure 5.10   Resistance management strategies to *Bt* transgenic plants.

survival of heterozygotes. The likelihood of these occurring needs to be determined experimentally before this strategy can be implemented. Monsanto requires growers commitment to set aside a certain portion for refugia - either 4% non-insecticide treated/non-*Bt* plants, or 20% insecticide treated, but not with *Bt* - and they must agree to one of these strategies (Sims, 1996).

5.4.1   Bt-*Cotton*

Introduction of Bollgard® cotton confers crop-resistance to some cotton pests via the expression of *Bt* endotoxin and has provided excellent control of tobacco budworm (*Heliothis virescens*). Nevertheless, the *Bt*-endotoxin is intrinsically less potent against the related cotton bollworm (*Heliothis zea*) and this appears to have led to some instances of bollworm damage in cotton not otherwise treated. An over-spraying of *Bt*-cotton with the synthetic pyrethroid insecticide λ-cyhalothrin is reported to provide significant yield increases (Mink et al., 1997)). It was demonstrated in a laboratory study that insects that survived exposure to sprayable *Bt* products at generally sub-lethal doses were susceptible to λ-cyhalothrin than unexposed pests (Harris, 1998). This study concluded that oversprays of λ-cyhalothrin on *Bt*-cotton tended to reduce the level of surviving lepidopteran pests to very low levels. This vastly enhanced the dilution effect offered by the influx of susceptible larvae from refugia and thus worked in concert to offer a sustainable resistance management strategy (Broadhurst, 1998).

5.4.2   Bt-*Corn*

Resistance management in *Bt* corn is currently based on two complementary principles: high dose and refuge. *Bt* corn is designed to produce very high levels of *Bt* proteins, much higher than levels found in *Bt* insecticides (Ostlie et al., 1997). The intent is to kill all European corn borer (ECB) larvae with no genes for resistance, plus those with one copy of a resistance gene. In any population of ECB, a few borers will have two copies of genes conferring resistance (rr), some will have one copy of the gene (rs) and most will have none (ss). The assumption inherent in this resistance management approach is that *Bt* hybrids have achieved the high-dose objective. If the high-dose objective is not achieved, then corn borer larvae with one copy of a resistance gene may survive to adulthood and mate with other resistant moths. Most of the offspring from these matings would be resistant to *Bt* corn. The second principle of the resistance management plan is the use of refuges. The purpose is to provide a source of ECBs, not exposed to *Bt* corn or *Bt* insecticides, to mate with potential resistant moths emerging from nearby *Bt* corn. The goal is to produce an overwhelming number of susceptible moths to every resistant moth. A refuge is any non-*Bt* host, including non-*Bt* corn, potatoes, sweet corn, cotton or native weeds that occur near *Bt* corn. Based on current knowledge of European corn borer biology, pesticide resistance studies and computer simulation models, the amount of refuge required has been estimated. It is approximately 20-30% of the corn acreage, so that 20-30% larval population is prevented from *Bt* protein exposure. This assumes that corn borers from alternative hosts emerge at similar times as corn borers from *Bt* corn.

5.4.3   Bt-*Potato*

A resistance management program for transgenic potato, NewLeaf®, prepared by the developing company Monsanto has been accepted by the U.S. Environmental Protection Agency (Stone and Feldman, 1995). This includes, a) incorporation of NewLeaf potatoes into IPM programs as an integral and not stand-alone component, b) monitoring of Colorado potato beetle (CPB) populations for susceptibility to Cry3A protein after establishing a base-line susceptibility, c) high dose expression of Cry3A in potatoes to control CPB heterozygous for resistance alleles, leaving only homozygous resistant beetles with a ability to survive NewLeaf plants, d) refugia, as host for susceptible insects through non-transgenic potatoes and e) search and development of Cry3A proteins with a distinct mode of action, use of multiple gene and alternate gene strategies with a potential for substantially delaying or halting resistance. Initially, Monsanto voluntarily placed a requirement on growers that no more than 80% of their crop could be planted with the transgenic potato. In 1999, EPA made it mandatory for plantation of 20% structured refuge of non-Bt potatoes in close proximity to the Bt potato field, as part of the registration (EPA, 1999).

5.5 Bt *Resistance Genes and Estimation of Resistance Development*

Tabashnik et al. (1997) studied diamondback moth (*Plutella xylostella*) resistance to four *Bt* toxins, *i.e* Cry1Aa, Cry1Ab, Cry1Ac, and Cry1F. They found that a single autosomal recessive gene conferred extremely high resistance to all the four *Bt*-endotoxins. This finding was significant, as Cry1 toxins are reported (Chambers et al., 1991) to have diverse insecticidal activity spectra (Table 5.3), even though they have greater than 70% sequence homology. This precludes the possibility of employing any of these single Cry1 proteins applications by rotation for delaying resistance development. As Bollgard cotton contains Cry1Ac and transgenic *Bt* corn contains Cry1Ab respectively, some concerns have been raised that multiple resistance genes found in diamondback

Table 5.3   Insecticidal activity spectra of different Cry1 protein toxins (Adapted from Carlton and Gawron-Burke, 1993)

| Cry1 protein | LC_{50} (ng Cry1/mm² diet surface) | | | |
	HV[a]	HZ[b]	ON[c]	SE[d]
Cry1Ab	0.68	16.9	0.27	38.8
Cry1Ac	0.04	1.4	0.11	>57
Cry1F	0.31	>57	0.27	25.6

(a) *Heliothis virescens* (tobacco budworm), (b) *Helicoverpa zea* (corn earworm), (c) *Ostrinia nubilalis* (European corn borer) and (d) *Spodoptera exigua* (beet armyworm).

moth may not be specific to these species only and that these genes may be present in other lepidopteran species as well. The implication could be that a polyphagous (insects that feed on many crops) lepidopteran larvae such as cotton bollworm (*Helicoverpa zea*), that develops resistance to Cry1Ac by feeding on Bollgard cotton could become quickly resistant to Cry1Ab used in transgenic corn or vice versa. These authors also reported that 21% of individuals from a susceptible strain were heterozygous for multi-toxin resistance gene that meant that resistance allele frequency was 10 times higher than previously established. This suggested that resistance might evolve to same groups of toxins much faster than previously expected. However, Roush and Shelton (1997) observed that the rate of evolution of resistance is much more sensitive to changes in selection than in the initial resistance allele frequency. Therefore it is difficult to arrive at a reliable initial resistance allele frequency in field population from the rates of resistance development in the field.

Gould et al. (1997) provided the first direct estimate of the initial frequency of *Bt* resistance genes in susceptible field populations of a major cotton pest, the tobacco budworm, *Heliothis virescens*. Resistant individuals are rare initially, and it is inherently difficult to estimate their frequency before populations are exposed to an insecticide. The frequency results from a balance between creation of resistant genotypes by mutation and selection against such mutants when the insecticide is not present. Thus, 2000 male insects from four cotton-growing areas were individually crossed with females of a strain selected for its extreme high resistance to Cry1Ac, the *Bt* gene used in cotton against tobacco budworm. Three males from the sample of 1025 successful crosses were found to be carrying an allele for resistance to *Bt* toxin, that led Gould and coworkers to estimate the field frequency of *Bt* resistance alleles to about 3 in 2000. This estimate is considerably higher than those assumed in earlier theoretical models and thus forebodes that resistance will evolve rapidly if pest populations receive prolonged and uniform exposure to *Bt* toxins. This also implied that the success of the first generation of *Bt*-transgenic plants could be short lived. Nevertheless, Gould et al. noted that if the current resistance management mandates (4% refuge) are followed, it should take at least 10 years before resistance becomes a problem in *Heliothis virescens*, because of the mortality of heterozygotes. This would not only outlast most intensive use of *Bt* sprays on the diamondback moth, but also the historical average of chemical sprays against cotton bollworms.

6   Concluding Remarks

Many insect populations have been shown to evolve resistance to *Bt* δ–endotoxins in the laboratory, although only *P. xylostella* has developed high level resistance in the field. The mode of resistance to *Bt* in *P. xylostella* is characterised by more than 500-fold resistance to at least one Cry1A toxin, recessive inheritance, little or no cross-resistance to Cry1C and reduced binding of at least one Cry1A toxin (Tabashnik et al., 1998). Analysis of resistance to *Bt* in the diamondback moth and two other species (tobacco budworm and Indian meal moth) suggested that this mode of resistance was common, although it may not be the only means by which insects attained resistance to *Bt*.

   Bt-based insecticides are frequently used in intensive agriculture, either in conjunction with conventional insecticides as an Integrated Pest Management strategy or as an alternative to synthetic insecticides against which resistance has occurred. It is anticipated that the future will see a better balance to transgenic, biological and chemical means of insect control that reduces the selection pressure against any one product. In this regard, the judicious use of chemicals in combination with *Bt*-crops is likely to extend the effectiveness of this technology in transgenic plants. It is also anticipated that improved understanding of the complex interplay among *Bt* toxins, their bacterial hosts, their target organisms and the ecosystems they share will allow for the long-term, effective use of *Bt* toxins for pest management.

References

1.   Bauer, L. S., 1995. "Resistance: A threat to the insecticidal crystal proteins of *Bacillus thuringiensis*", *Florida Entomol.*, 78(3), 414-443.

2.   Bosch, D., B. Schipper, H. van der Kleij, R. A. de Maagd and W. J. Stiekema, 1994. "Recombinant *Bacillus thuringiensis* crystal proteins with new properties: possibilities for resistance management", *Bio/Technol.*, 12, 915-918.

3.   Broadhurst, M. D., 1998. "The influence of the molecular basis of resistance in insecticide discovery", *Phil. Trans. R. Soc. Lond.* B, 353, 1723-1728.

4.   Carlton, B. C., (Ecogen Inc.), 1995. "Developing new recombinant strains of *Bt*", Information Systems for Biotechnology/ NBIAP News Report, Special

Issue on *Bt* – December 1995, http://www.nbiap.vt.edu/news/1995/ news95. dec.html dec9506.

5.   Carlton, B. C., and C. Gawron-Burke, 1993. "Genetic improvement of *Bacillus thuringiensis* for bioinsecticide development", in *Advanced Engineered Pesticides*, Leo Kim, ed., Marcel Dekker, New York, NY, pp. 43-61.

6.   Chambers, J. A., A. Jelen, M. P. Gilbert, C. S. Jany, T. B. Johnson and C. Gawron-Burke, 1991. "Isolation and characterization of novel insecticidal crystal protein gene from *Bacillus thuringiensis* subsp. *aizawai*", *J. Bacteriol.*, 173, 3966-3976.

7.   Crickmore, N., E. J. Bone, J. A. Williams and D. J. Ellar, 1995. "Contribution of the individual components of the δ-endotoxin crystal to the mosquitocidal activity of *Bti*", *FEMS Microbiol Letts.*, 131(3), 249-254.

8.   DuBois, N. R. and D. H. Dean, 1995. "Synergism between Cry IA insecticidal crystal proteins and spores of *Bacillus thuringiensis*, other bacterial spores, and vegetative cells against *Lymantria dispar* (Lepidoptera: Lymantriidae) larvae", *Environ. Entomol.*, 24 (6), 1741-1747.

9.   EPA Office of Pesticide Programs, 1999. "EPA and USDA position paper on insect resistance management in Bt crops", http://www.epa.gov/pesticides/ biopesticides/otherdocs/bt_position_paper_618.htm.

10.   Ferre, J., M. D. Real, J. Van Rie, S. Jansens and M. Perferoen, 1991. "Resistance to the *Bacillus thuringiensis* bioinsecticide in a field population of *Plutella xylostella* is due to a change in a midgut membrane receptor", *Proc. Natl. Acad. Sci., USA*, 88, 5110-5123.

11.   Georghiou, G. P., and A. Lagunes-Tejeda, 1991. "The occurrence of resistance to pesticides in arthropodes", FAO, Rome, Italy.

12.   Georghiou, G. P. and M.C. Wirth, 1997. "Influence of exposure to single versus multiple toxins of *Bti* on development of resistance in the mosquito *Culex quinquefasciatus*", *Appl. Environ. Microbiol.* 63(3), 1095-1101.

13.   Glick, B. R. and J. J. Paternak, 1998. "Microbial Insecticides", in *Molecular Biotechnology: Principles and Applications of Recombinant DNA*, 2nd ed., ASM Press, washington, DC, pp. 377-398.

14.   Gould, F., A. Anderson, A. Jones, D. Sumerford, D. Heckel, J. Lopez, S. Micinski, R. Leonard and M. Laster, 1997. "Initial frequency of alleles for resistance to *Bt*-toxins in field populations of *Heliothis virescens*", *Proc. Natl. Acad. Sci.,U.S.A.*, 94, 3519-3523.

15.   Gould, F., A. Martinez-Ramirez, A. Anderson, J. Ferre, F. J. Silva and W. A. Moar, 1992. "Broad-spectrum resistance to *Bacillus thuringiensis* toxins in Heliothis virescens", *Proc. Natl. Acad. Sci. USA*, 89, 7986-7990.

16.   Harris, J. G., 1998. "The usage of Karate® (λ-cyhalothrin) over-sprays in combination with refugia, as a viable and sustainable resistance management strategy for *Bt*-cotton", *Proc. Beltwide Cotton Production Conf.*, Memphis, TN, National Cotton Council, pp. 1217-1221.

17.   Lambert, B. and M. Peferoen, 1992. "Insecticidal promise of *Bacillus thuringiensis*", *BioScience*, 42(2), 112-121.

18.   Lee, M. K., A. Curtiss, E. Alcantara and H. D. Dean, 1996. "Synergistic effect of the *Bacillus thuringiensis* toxin CryIAa and CryIAc on the gypsy moth, *Lymantria dispar*", *Appl. Environ. Microbiol.*, 62 (2), 583-586.

19.   Lee, M. K., F. Rajamohan, F. Gould, and D. H. Dean, 1995. "Resitance to
       Bacillus thuringiensis Cry1A δ-endotoxins in a laboratory *selected Heliothis
       virescens* strain is related to receptor alteration", *Appl. Environ. Microbiol.*, 61,
       3836-3842.

20.   Liu, Y-B, B. E. Tabashnik, M. Pusztai-Carey, 1996. "Field evolved resistance
       to *Bacillus thuringiensis* toxin CryIC in diamondback moth (*Lepidoptera:
       Plutellidae)*", *J. Econ. Entomol.*, 89(4), 798-804.

21.   Liu, Y-B and B. E. Tabashnik, 1997. " Experimental evidence that refuges
       delay insect adaptation to *Bacillus thuringiensis*", *Proc. R. Soc. Lond.* B, 264,
       605-610.

22.   Liu, Y-B, B. E. Tabashnik, W. J. Moar and R. A. Smith, 1998. "Synergism
       between *Bacillus thuringiensis* spores and toxins against resistant and
       susceptible diamondback moth (*Plutella xylostella*)", *Appl. Environ.
       Microbiol.*, 64, 1385-1389.

23.   Marrone, P. G. and S. C. Macintosh, 1993. "Resistance to *Bacillus
       thuringiensis* and resistance management", in *Bacillus thuringiensis, An
       Environmental Biopesticide: Theory and practice*, P. F. Entwistle and, J. S.
       Cory, M. J. Bailey and S. Higgs, eds., John Wiley, pp. 221-235.

24.   McGaughey, W. H. and M. E. Whalon, 1992. "Managing insect resistance to
       Bacillus thuringiensis toxins", *Science*, 258, 1451-1455.

25.   McGaughey, W. H. and R. W. Beeman, 1988. "Resistance to *Bacillus
       thuringiensis* in colonies if Indianmeal moth and almond moth (Lepidoptera:
       Pyralidae)", *J. Econ. Entomol.*, 81, 28-33.

26.   McGaughey, W. H., F. Gould and W. Gelernter, 1998. "*Bt* resistance
       management", *Nature Biotechnol.*, 16, 144-146.

27.   McGaughey, W. H., 1985. "Insect resistance to the biological insecticide
       *Bacillus thuringi*ensis", *Science*, 229, 193-195.

28.   Mink, J. H. and S. Martin, 1997. "Performance and benefits of Karate
       insecticide on Bollgard cotton", *Proc. Beltwide Cotton Prodn. Conf.*, Memphis,
       TN, National Cotton Council, pp.898-899.

29.   Moar, W. J., M. Pusztai-Carey and M. J. Adang, 1995. "Development of
       Bacillus thuringiensis CryIC resistance by *Spodoptera exigua* (Hubner)
       (Lepidoptera: Noctuidae)", *Appl. Environ. Microbiol.*, 61, 2086-2092.

30.   Muller-Cohn, J., J. Chaufauk, C Buisson, N. Gilois, V. Sanchis and D.
       Lereclus, 1996. "*Spodoptera littoralis* (Lepidoptera: Noctudae) resistance to
       CryIC and cross-resistance to other *Bacillus thuringiensis* crystal toxins", *J.
       Econ. Entomol.*, 89(4), 791-797.

31.   Menn, J. J., 1996. "Biopesticides: Has their time come?", *J. Environ. Sci.
       Health*, B31(3), 383-389.

32.   Nielsen-Leroux, C., J. F. Charles, I. Thiery and G. P. Georghiou, 1995.
       "Resistance in a laboratory population of *Culex quinquefasciatus* to *Bacillus
       sphaericus* binary toxin is due to change in the receptor on midgut brush border
       membrane", *Eur. J. Biochem,.* 228, 206-210.

33.   Oppert, B., K. J. Kramer, R. W. Beeman, D. Johnson and W. H. McGaughey,
       1997. "Proteinase-mediated resistance to *Bacillus thuringiensis* insecticidal
       toxins", *J. Biol. Chem.*, 272, 23473-23476.

34.   Ostlie, K. R., W. D. Hutchison and R. L. Hellmich, 1997. "*Bt* corn and European corn borer", *NCR Publication 602*. University of Minnesota, St. Paul, MN

35.   Poncet, S., A. Delecluse, A. Klier and G. Rapoport, 1995. "Evaluations of synergistic interactions among the CryIVA, CryIVB and CryIVD toxic components of *Bti* crystals", *J. Invertebr. Pathol.*, 66(2), 131-135.

36.   Perez, C. J., A. M. Shelton and R. T. Roush, 1997. "Managing diamonback moth (Lepidoptera: Plutellidea) resistance to foliar applications of *Bacillus thuringiensis*: Testing strategies in field cages", *J. Econ. Entomol.*, 90, 1462-1470.

37.   Rodcharoen, J. and M. S. Mulla, 1995. "Comparative ingestion rates of *Culex quinquefasciatus* susceptible and resistant to *Bacillus sphaericus*", *J. Invertebr. Pathol.*, 66, 242-248.

38.   Roush, R. T. and A. M. Shelton, 1997. "Assessing the odds: The emergence of resistance to *Bt* transgenic plants", *Nature Biotechnol.*, 15, 816-817.

39.   Schnepf, E., N. Crickmore, J. Van Rie, D. Lereclus, J. Baum, J, Feitelson, D. R. Zeigler and D. H. Dean, 1998. "*Bacillus thuringiensis* and its pesticidal crystal proteins", *Microbiol. Molecular Biol. Revs.*, 775-806.

40.   Sims, S. B., J. T. Greenplate, T. B. Stone, M. A. Caprio and F. L. Gould, 1996. "Monitoring strategies for early detection of *Lepidoptera* resistance to *Bt*-ICPs, *Molecular Genetics and Evolution of Pesticide Resistance*", American Chemical Society, Washington, D.C., pp. 229-247.

41.   Sims, S. (Monsanto), 1996. *Conference on Biological Control*, Cornell Univ., April 11-13, 1996, http://www.nysacs.cornell.edu/ent/bcconf/talks/ industry_q_a-.html.

42.   Stone, T. and J. Feldman, 1995. "Development of a comprehensive resistance management plan for NewLeaf potatoes", *Information Systems for Biotechnology/NBIAP News Report*, Special Issue on *Bt*, December 1995,http://www.nbiap.vt.edu/news/1995/news95.dec.html.dec9506.

43.   Stone, T. B., S. R. Sims and P. G. Marrone, 1989. "Selection of tobacco budworm for resistance to a genetically engineered *Pseudomonas fluorescens* containing the δ-endotoxin of *Bacillus thuringiensis* subsp. *kurstaki*", *J. Invertibr. Pathol.*, 53, 228-234.

44.   Tabashnik, B. E., 1994. "Evolution of Resistance to *Bacillus thuringiensis*", *Annu. Rev. Entomol.*, 39, 47-79.

45.   Tabashnik, B. E., F. R. Groeters, N. Finson, Y-B liu, M. W. Johnson, D. G. Heckel, K. Luo and M. S. Adang, 1996. "Resistance to *Bacillus thuringiensis* in *Plutella xylostella*: the moth heard round the world", in *Molecular Genetics and Evolution of Pesticide Resistance*, American Chemical Society, Washington, D.C., pp. 130-140.

46.   Tabashnik, B. E., Y-B Liu, N. Finson, L. Masson and D. G. Heckel, 1997. "One gene in diamondback moth confers resistance to four *Bt*-toxins", *Proc. Natl. Acad. Sci. , U.S.A.*, 94(5), 1640-1644.

47.   Tabashnik, B. E. Y-B Liu, T. Malvar, D.G. Hackel, L. Masson and J. Ferre', 1998. "Insect resistance to *Bacillus thuringiensis*: uniform or diverse", *Phil. Trans. R. Soc. Lond.* B, 353, 1751-1756.

48.   Tang, J. D., A. M. Shelton, J. Van Rie, S. deRock, W. J. Moar, R. T. Roush,
       and M. Peferoen, 1996. "Toxicity of *Bacillus thuringiensis* spore and crystal
       protein to resistant diamondback moth (*Plutella xylostella)*", *Appl. Environ.
       Microbiol., 62* (2), 564-569.
49.   Van Rie, J., W. H. McGaughey, D. E. Johnson, B. D. Barnett and H. Van
       Mallaert, 1990. "Mechanism of insect resistance to the microbial insecticide
       Bacillus thuringiensis", *Science*, 247, 72-74.
50.   Whalon, M. E., D. L. Miller, R. M. Hollingworth, E. J. Grafius and J. R. Miller,
       1993. "Selection of a Colorado potato beetle (Coleoptetra: Chrysomelidae)
       stain resistant to *Bacillus thuringiensis*", *J. Econ. Entomol.*, 86, 226-233.
51.   Wirth, M. C., G. P. Georghiou and B. A. Federici, 1997. "CytA enables CryIV
       endotoxins of *Bacillus thuringiensis* to overcome high levels of Cry IV
       resistance in the mosquito, *Culex quiquefasciatus*", *Proc. Natl. Acad. Sci.,
       USA*, 94, 10536 – 10540.

Part II

Viral Insecticides, Biofungicides, Bioherbicides, and Mycoinsecticides

6

Natural and Recombinant Viral Insecticides

1   Introduction

Demand for effective, more selective, safer pesticides, consistent with sustainable agriculture continues to push the search for newer systems as well as modifying the existing ones. Insects are subject to diseases caused by viruses, bacteria, fungi, protozoans and nematodes. Many members of these groups have been evaluated as biocontrol agents. Most pathogens and nematodes tend to infect primarily insect larvae and are only effective against this stage. Although bacterial control agents of *Bacillus thuringiensis* family are currently the leading bioinsecticides, viral and fungal agents are gradually making headway. For example, baculoviruses directly compete with *Bt* for the control of lepidopterous insects, albeit feebly at present. With the advances in development of recombinant baculoviruses, that are more virulent and fast acting, baculoviruses have a great potential for effective integration into pest management system.

2   Baculoviruses

The baculoviruses are the naturally occurring invertebrate-specific pathogens that infect some important lepidopteran (butterflies and moths) hymenopteran (forest pests) insects. Their high pathogenicity, narrow host range and complete safety to vertebrates and plants, make them ideal candidates for biological control of these insects. The viral pesticides can be applied using conventional techniques and do not create the problem associated with residues. Baculoviruses infect predominantly holomentabolons insects and almost all pest species within the Lepidoptera are susceptible to infection by at least one of the baculoviruses. Despite their potential, viral insecticides are employed much less than they could be in crops and forests, due to difficulty in virus stability and, most importantly, slower speed of action than that achieved with chemical pesticides.

Amongst the best examples involving use of baculoviruses is the exceptional control of velvetbean caterpillar (*Anticarsia gemmatalis*) on soybean in Brazil and coconut rhinoceros beetle (*Oryctes rhinoceros*) in Andaman Islands in India. The *Anticarsia gemmatalis* nuclear polyhedro virus (AgNPV) is highly virulent to host larvae at a low dose (1.5 x 10^{11} OB/ha) and it is efficiently transmitted in the host population by natural enemies and abiotic factors. As a result, one AgNPV application per cropping season is sufficient to control the insect pest and is as effective as two insecticide applications in a season. As a leaf feeder, *A. gemmatalis* is exposed to applied AgNPV and usually no other key insect occurs simultaneously on the crop. Also, soybean tolerates considerable defoliation with no reduction in yield (Moscardi, 1999). Another successful application of baculovirus has been the control of coconut rhinoceros beetle (*O. rhinoceros*), a serious pest that threatens coconut production. The application of baculovirus effectively controls this pest at a low cost. The strategy employed in the Andaman Islands in India, has been to induce virus epidemics in beetle populations, in the newly invaded areas, by infecting and liberating beetles with *Oryctes* baculovirus. Results comparing levels of beetle damage to coconut, before and after virus introductions, have been 80-90% reductions (Jacob, 1996). Once introduced, the virus seems to maintain itself adequately without additional intervention.

A number of recent reviews on naturally occurring baculoviruses have appeared in the literature (Federici, 1999; Moscardi, 1999; Volkman, 1997; Hammer et al., 1995; Cunningham, 1995; and Crook and Jarret, 1991). The developmental efforts on recombinant baculoviruses have also been reviewed (Treacy, 1999; Possee, 1997; Bonning and Hammock, 1992, 1996; Hoover et al., 1996; Maeda, 1995; and Wood, 1995). An overview of the progress achieved in realizing the potential of natural and recombinant baculoviruses has been provided in this chapter.

2.1 Classification

The baculoviruses are divided into two sub-groups based on their morphology. In sub-group A, Nuclear Polyhedrosis Viruses (NPVs) have many virus particles embedded in each occlusion body, which are called polyhedra because of their shape. Some NPVs have a single nucleocapsid within each virus particle and are designated SNPVs. In contrast the MNPVs have multiple nucleo-capsids within each virus particle. These have been isolated from several hundred insect species including *Spodoptera littoralis*, *S. frugiperda*, *Lymantria dispar*, *Helicoverpa zea* and *Heliothis armigera*. The NPVs are named after the insect species from which they were first isolated, followed by the appropriate baculovirus sub-group name.

In sub-group B, Granulosis viruses (GVs) have a single virus particle embedded in each occlusion body, which is capsule-shaped and called granules. Granulosis virus (GV), does not grow well in cell culture, thus interest in their industrial exploitation has been negligible. However, GVs can be used against a number of pest species, including *Pieris brassicae* on cabbages and *Plodia interpunctella*, a stored product pest.

2.2 Characteristics

The baculoviruses are characterized by the presence of rodshaped nucleocapsids, which are further surrounded by a lipoprotein envelope (Figure 6.1) to form the

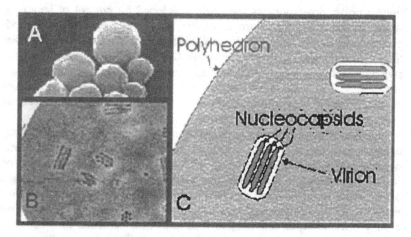

Figure 6.1 Electron micrographs of (a) Baculovirus particles or polyhedra
(b) Cross-section of polyhedra and (c) Diagram of polyhedra
cross-section (Courtesy Jean Adams and V. D'Amico)

virus particles. Certain baculoviruses, specifically nuclear polyhedrosis viruses (NPVs) have a unique life cycle that involves the temporally regulated expression of two functionally and morphologically different viral forms, the budded form and the occluded form. Nuclear polyhedrosis viruses produce large polyhedral occlusion bodies, which contain enveloped virus particles within the nucleus of an infected cell. The occlusion body is composed of a matrix comprising a 29 kDa protein known as polyhedron. These occlusion bodies serve to protect the virus particles in the environment and also provide a means of delivering the virus particles to the primary site of infection in insects, the midgut epithelial cells. The budded form of the virus is involved in the secondary infection, virtually affecting all the tissues within the host larvae. Replication of the virus in other organs creates extensive tissue damage and eventually death. When an infested insect dies, millions of polyhedra are released. This allows the virus to persist in the environment although its host insect larvae may be present for only a few weeks during the year.

The pathogenicity of baculoviruses has been characterized mostly by their median lethal dose (LD_{50}), an estimate of infectivity of the virus to its host. Virulence, which in the case of baculoviruses is best determined by the speed by which a given virus elicits the desired response, is an important measure of the effectiveness of the viruses as a biopesticidal agent (Van Beek and Hughes, 1998).

2.3   Nuclear polyhedrosis viruses (NPVs) - Potential as Viral Insecticides

Nuclear polyhedrosis viruses are being developed for control of lepidopterous larvae. In the USA and Europe, a few baculovirus products are produced commercially for use in field crops (Table 6.1). The companies like Dupont, Biosys (now Thermo Trilogy), American Cyanamid and Agrivirion (to name a few) have active research programs for development of agricultural-use viral insecticides. For example, Biosys had introduced two baculovirus-based products, Spod-X for control of beet armyworm and Gemstar LC for control of tobacco budworm and cotton bollworm.

The limited commercial penetration of the market by viral insecticides suggests that major factors constrain their wider use. Most importantly, baculoviruses compete inadequately with classical insecticides because of the viruses' slow speed of kill. The time period between the initial infection by baculoviruses to the death of diseased larvae varies from 2/3 days to 2/3 weeks, depending on the virulence of virus isolate and nutrition of larval host. Further, an NPV-infected larva continues to feed until time of death; thus continues to damage the crop after treatment. Therefore most commercially successful baculoviruses have been used primarily on crops which can sustain damage without major economic losses, such as forests (Possee et al., 1997). Some of

Table 6.1 Commercial baculovirus formulations for use in field crops

Product Name (Manufacturer)	Baculovirus	Pests/Crops
Spod-X (Thermo Trilogy)	Spodoptera exigua Nucleopolyhedro Virus (SeNPV)	Beet armyworm of floral crops
GemStar (ThermoTrilogy), Elcar (Novartis)	Helicoverpa zea Nucleopolyhedro Virus (HzNPV)	Several Heliothis/ Helicoverpa spp. on cotton, corn, tobacco, sorghum, soybean and tomato
Cyd-X, (Thermo Trilogy), Madex (Andermatt Biocontrol, Switzerland), Granusal (Behring AG, Werke, Germany), Caprovirusine (NPP, France)	Cydia pomonella Granulo Virus (CpGV)	Codling moth in walnuts, apples, pear and plum.
Mamestrin (National Plant Protection, France)	Mamestra brassicae Nucleopolyhedro Virus (MbNPV)	Cabbage moth, American bollworm, diamondback moth.
VPN (Agricola El Sol, Brazil)	Anticarsia gemmatalis Nucleopolyhedro Virus (AgNPV)	Velvetbean caterpillar on soybean
Gusano (Thermo Trilogy)	Autographa californica Nucleopolyhedro Virus (AcNPV)	Alphalpha looper
Spodopterin (National Plant Protection, France)	Spodoptera littoralis Nucleopolyhedro Virus (SlNPV)	Cotton and vegetables

the baculovirus products for control of forest pests have been registered by The U. S. Department of Agriculture Forest Service and Canadian Forest service (Table 6.2).

In most cases, it is difficult to develop cell cultures which will support replication of viruses, particularly that of baculoviruses. However, there are two wide host range baculoviruses that are easy to work with and have good potential for being produced by fermentation. One of these is *Autographa californica* multiple nuclear polyhedro virus (AcMNPV), for which very good cell cultures have been available and therefore it is well studied. This virus, originally isolated from the alfalfa looper is of particular interest for insect pest management. The virus is known to infect 39 species of Lepidoptera, including

Table 6.2   Commercial baculovirus products for control of forest pests

Product	Virus	Target	Crop
Gypchek	*Lymantria Diaspora* Nucleopolyhedro Virus (LdNPV)	Gypsy moth	Broad-leaved trees
TM BioControl-1, Virtuss	Orgyia pseudotsugata Nucleopolyhedro Virus (OpNPV)	Tussock moth	Douglas fir
Virox	*Neodiprion sertifer* Nucleopoyhedro Virus (NeseNPV))	Spruce saw fly	Pine

those of the major pests *Heliothis*, *Spodoptera* and *Trichoplusia* (Cunningham, 1995).

Another multiple nucleopolyhedro baculovirus is that of the celery looper, *Anagrapha falcifera* (Kirby) (AfMNPV), that has also shown potential for development as a microbial insecticide. Much like AcMNPV, the AfMNPV has a broad host range and infects 31 species of Lepidoptera in 10 families (Hostetter and Putler, 1991). In a comparative field study, AfMNPV has shown greater virulence than *Heliothis* NPV toward *H. Zea* on corn ears. It was also found that AfMNPV application reduced damage and ear protection was comparable or better to *Btk* (Javelin®) and *Bt-aizawai* (XenTari®) products (Pingel and Lewis, 1997).

2.4   Mode of Action of Baculoviruses

Viral infection debilitates the host larvae resulting in reduction of development, feeding and mobility and increasing exposure to predation. On ingestion of polyhedra by insect larva, the polyhedra move to the midgut where the alkaline environment and possibly proteolytic action, dissolve the polyhedrin coat releasing the infectious nucleocapsids. The released nucleocapsids bind to the midgut epithelial cells, migrate through the cytoplasm and uncoat in the nucleus. Viral replication in the nucleus produces progeny nucleocapsids that bud through the plasma membrane of infested cells into the circulatory system of the insect. The open circulatory system of the insect provides the virus with access to other tissues within the host larvae. This results in secondary infection of tissues throughout the insect. Nucleocapsids may then be enclosed within the polyhedra in the nucleus, or continue to spread infection within the larval host, as shown in Figure 6.2.

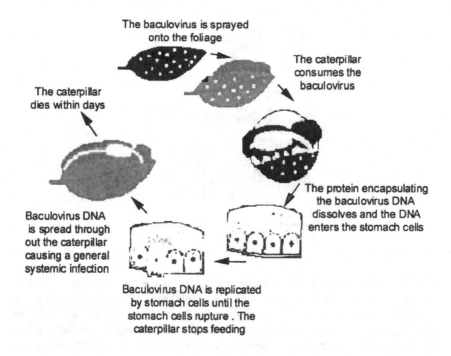

The baculovirus is sprayed onto the foliage

The caterpillar consumes the baculovirus

The caterpillar dies within days

The protein encapsulating the baculovirus DNA dissolves and the DNA enters the stomach cells

Baculovirus DNA is spread through out the caterpillar causing a general systemic infection

Baculovirus DNA is replicated by stomach cells until the stomach cells rupture . The caterpillar stops feeding

Figure 6.2   Mode of action of baculoviruses (Adapted from Georgis, 1996).

Post-larval effects may include lower pupal and adult weights, as well as reduced reproductive capacity and longevity. The polyhedra that are 1-10 µm in size account for up to 30% of the dry weight of the infected larvae. There may be up to 30 or more polyhedra in infected cells and late instar larvae may produce up to 10^{10} polyhedra before death. The insect may continue to feed till death occurs from viral infection, depending on the virus concerned and environmental factors. Infected defoliating larvae usually climb to the upper parts of the plants, dying in 5-8 days, depending on biotic and abiotic factors. Upon death, these are seen hanging limply from foliage and have a characteristic shiny-oily appearance, Figure 6.3.

Park et al. (1996) have studied the baculovirus *Lymantria dispar* NPV and found that it interferes with insect larval development by altering the host's hormonal system. The level of hemolymph ecdysteroids, the insect molting hormone, was higher in virus-infected larvae than in uninfected controls. There-

Figure 6.3   Virus killed caterpillar
(Courtesy Roger T. Zerillo, ref. 33).

fore, it is likely that the insects are no longer under the control of the hormonal
system after infection by viruses. Volkman (1997) observed that there is a fierce
competition at a molecular level between the baculovirus and its host over the
final disposition of the resources of the host i.e. whether they go toward building
polyhedra or a moth. The higher level of ecdysteroid prevents subsequent
conversion of larvae to moths, while at lower concentrations molting proceeds
normally.

2.5   *Production of Baculoviruses*

Viruses are facultative parasites and will only replicate in living cells, making it
necessary for these to be produced either in host insect larvae (*in vivo*) or in
susceptible cell culture (*in vitro*). Until recently, NPVs have been produced *in
vivo* and harvested from insect cadavers and processed as powders. This method
whilst adequate on a small scale, is costly to scale up and difficult to assess for
quality control in a commercial environment. Recent advances in tissue culture
technology and deep fermentation should enable cheaper production of NPVs.
Chakraborty et al. (1995) reported *in vitro* production of wild type *Heliothis*
baculoviruses. The viruses, including *Heliothis zea* and *Heliothis armigera*
NPVs in a *H. zea* cell line, adapted to grow in suspension culture using serum
free medium. Several companies are also pursuing this technology. In addition
Crop Genetics International (now Thermo Trilogy)) and AgriVirion (New York,
NY) have recently developed more cost effective *in vivo* production systems. It
is anticipated that these advances will significantly reduce the production costs
of viral pesticides.

2.6   Formulation of Baculoviruses

The majority of the baculoviruses are formulated as concentrated wettable powders. The corn earworm (*Helicoverpa zea*) NPV biocontrol product Elcar[^] (Novartis) is formulated by diluting with an inert carrier and spray drying or air-drying. Formulation of *Anticarsia gemmatalis* NPV with amorphous silica, attapulgite and kaolinite clays were found to maintain the activity after a year of storage, while in the presence of bentonite, activity declined to 50% in the same period (Medugno et al., 1997). *H. zea* NPV for control of *H. zea* and *H. virescens* in cotton is produced as a liquid concentrated formulation GemStar[^] by Thermo Trilogy. On the other hand, gypsy moth (*Lymantria dispar L.*) NPV are freeze-dried, either with a carbohydrate or by acetone precipitation, Additives are frequently put into tank mixes, which can improve the physical performance of the baculovirus spray deposit. A typical formulation consists of adjuvants-surface active agents such as wetting and dispersing agents, spreaders, stickers, sunlight screens, buffers and also gustatory stimulants (for information on gustatory stimulants, see Chapter 4).

The abiotic factor that is most responsible for inactivation of baculoviral occlusion bodies in natural environment, is sun light; particularly wavelengths of 290-320 nm can inactivate most viruses. Nucleopolyhedro viruses have been reported to be even more photolabile than other insect pathogens, such as *Bt*. Presence of free water, for example, dew deposits on foliage, can accelerate destruction of NPVs by sunlight. Recent advances in protecting NPVs from photo degradation through the adoption of sun screening agents and optical fluorescent brighteners has increased their utility. Several effective dyes, such as lissamine green, acridine yellow, alkali blue and mercurochrome have been used as photoprotectants (Shapiro, 1995). Optical brighteners also absorb UV light and have been shown to significantly reduce photodegradation of NPVs and enhance viral activity. The addition of stilbene oxide brighteners has been shown to increase the efficacy of the NPV of the gypsy moth (*Lymantria dispar*). These substances have enhanced the efficacy of other baculoviruses, including the NPVs of *H. zea, H. virescens, A. californica, S. exigua, T. ni, C. includans, A. falcifera* and *A. gemmatalis*. A similar enhancing influence of a fluorescent brightener Calcofluor white M2R on nuclear polyhedrosis virus of *Anagrapha falcifera* (AfMNPV) has been reported with reduced LT_{50} values on four noctuid pests (Vail et al., 1996). The possible mode of action for these compounds when combined with viruses in the insect gut, has been described (Washburn et al., 1998). The brighteners may block sloughing of infected primary target cells in the insect midgut and thus affect the host susceptibility to the baculoviruses.

2.7 Enhancement of Viral Pesticidal Activity by Enhancin

The discovery of NPV enhancing proteins as synergists for NPVs adds another
dimension to the utility of NPVs as insect control agents. A 104 kDa protein
called Enhancin (formerly known as viral enhancement factor, VEF) contained
in the granules of occluded *T. ni* GV, increases the viral pesticidal activity in
noctuid moths at levels of 50 pico gram per insect. This protein appears to
comprise about 5% of TnGV polyhedra (Gallo et al., 1991). The biochemical
activity, held within the polyhedron protein matrix and released during the
alkaline dissolution of polyhedra in the insect midgut causes specific
biochemical and structural changes in peritrophic membrane (the interior lining
of the insect's intestine). The peritrophic membrane, which is made mostly of
chitin and proteins, is an important component of the insect's immune system
and serves as a protective barrier that separates the insect midgut from the
infection susceptible midgut epithelical cells. Enhancin binds to a major
mucinous protein of peritrophic matrix, destroying the matrix structure, thus
allowing the baculovirus to pass through the relatively impermeable matrix
unimpeded, resulting in the rapid death of the insect (Wang and Granados, 1997)
(Figure 6.4).

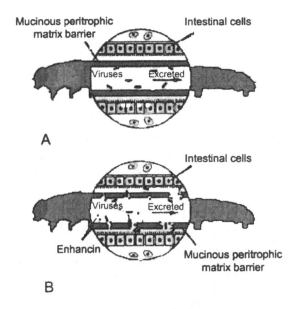

Figure 6.4   (A) Viruses cannot pass through intestine wall to kill insect.
(B) Enhancin protein attacks the immune system allowing
viruses to pass through (Courtesy: Granados, 1997).

In dose-response assays with *T. ni* larve, inoculated with AcMNPV, TnSNPV and *Anticarsia gemmatalis* MNPV, the addition of enhancin resulted in 3- to 16-fold reduction in $LD_{50}s$. Thus, the addition of supplemental enhancin in baculovirus pesticide products could significantly reduce application rates and would be particularly useful for the control of late instar larvae, that typically require very high application rates to catch infection.

Menn (1996) has also quoted unpublished research works suggesting that NPVs could be synergistically combined with reduced rate of conventional insecticides for insect control (for detailed discussion see Chapter 10).

3 Genetically Modified Baculoviruses for Insect Control

Biotechnology, through recombinant DNA technology has provided the means of overcoming some of the shortcomings of naturally ocurring baculoviruses, while maintaining or enhancing their desirable pest-specific characteristics. Nucleopolyhedro viruses have been genetically altered to enhance the speed with which they kill the target pest. This has been achieved through genetic manipulation of baculoviruses by exchange of genetic material between different baculoviruses or insertion of foreign genes into baculovirus genomes.

Baculoviruses are rod-shaped particles containing at least 25 different polypeptides and a large (100-200 kbp) circular double stranded DNA genome. The NPV double-stranded DNA is highly amenable for DNA recombinant technology through the use of the polyhedrin promoter gene that is suited for expression of exogenous genes such as neurotoxins, neuropeptides, juvenile hormone esterase and others.

Two major baculovirus expression systems have been developed for the production of recombinant proteins. These are based on nucleopolyhedro viruses derived from alfalfa looper (*Autographa californica*) and a similar virus from the silkworm (*Bombyx mori*). The entire sequences for the genome of both *Autographa californica* NPV (AcNPV) and *Bombyx mori* NPV (BmNPV) have been determined. The genome of AcNPV has a double-stranded covalently closed circular DNA of 134 kilobase (kb) pairs.

Recombinant baculoviruses are constructed in two stages due to the difficulty of manipulating the large genome directly. The foreign gene is incorporated initially into a baculovirus transfer vector. The gene-inserted plasmid is then propagated in the bacterium *Escherichia coli*. Most transfer vectors used are bacterial plasmid University of California (pUC) derivatives, which encode an origin of replication for propagation in *E. coli* and an ampicillin-resistance gene. The pUC fragment is ligated to a small segment of DNA taken from the viral genome. The foreign gene sequence is incorporated into a cloning site downstream of the promoter selected to drive expression. For

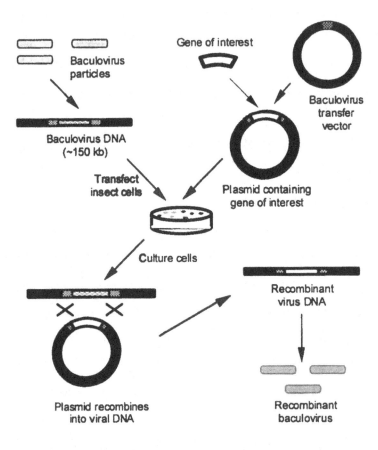

Figure 6.5   Construction of recombinant baculovirus

the second step, the transfer vector is mixed with DNA from the wild-type baculovirus. The engineered DNA is incorporated into the virus via homologous recombination events within the nucleus of cultured insect cells (Figure 6.5). The baculovirus system allows the precise insertion of foreign DNA without disruption of other genes, unlike genetic engineering in plants, which results in a rather random incorporation of new DNA into the genome (Bonning and Hammock, 1996).

Any gene coding for a protein that disrupts normal larval development or behavior and reduces feeding damage caused by the insect is a candidate for expression by a recombinant baculovirus for insect control program. Production during the infection of proteins that interferes specifically with insect metabolism or metamorphosis such as hormones, enzymes and toxins, might enhance the pathogenicity of these viruses. The production of a diuretic hormone, a mite toxin, a scorpion toxin or a modified juvenile hormone esterase enhances biological activity of baculoviruses and so these are the first examples of genetically improved baculovirus insecticides (Table 6.3).

The recombinant virus expresses the genes for rapid kill of the host and is subjected to strong negative selection pressure. A quicker death almost certainly would lead to fewer viral progeny. Resultantly, recombinant virus can not compete with its wild type counterpart in the field. Recombinant baculoviruses, therefore are being developed as biological insecticides for repeated application rather than as biological control agents that establish and recycle in the field.

3.1   Recombinant AcNPV Expressing Toxins

Arthropod venoms offer a rich source of insect-selective toxins. In fact, expression of insect selective toxins in the baculovirus expression system has proved to be highly successful for increasing virus efficacy in insect pest control. Baculoviruses expressing the insect specific toxin AaIT derived from the Algerian scorpion (Androctonus australis) are among the most promising recombinant baculoviruses developed yet. This neurotoxin which does not have any effect on mice, disrupts the flow of sodium ions in neurons of targeted insects, causing repetitive firing of motor nerves and overstimulation of skeletal muscle. Larvae infected with these recombinant baculoviruses exhibit dorsal arching and increased irritability and cessation of feeding. Lethal times are reduced by 25-40% compared with those of the wild-type virus. The feeding

Table 6.3   Genes that increase the insecticidal effects of baculoviruses
(Adapted from Maeda, 1995)

Name of Gene	Effect(s) of expressed gene on host insect	% increase in insect killing
Diuretic hormone	Disrupts water balance	20%
Juvenile hormone esterase	Feeding cessation (1st instar) and blackening	-
Bt toxin	Feeding cessation	$2 \times LD_{50}$
Scorpion toxin (AaIT)	Paralysis, body tremors and dorsal arching	30-40%
Mite toxin (Txp-1)	Paralysis	30-40%
Wasp toxin (Antigen 5)	Premature melanization and low weight gain	-

damage by *T. ni* larvae infected with recombinant virus is reduced by 50% on cabbage compared with damage caused by larvae infected with wild-type viruses. When AcNPV.AaIT was tested under field conditions, it was found to be even more effective. In the field, it killed the insect pests faster, decreased the damage to cabbage plants and reduced the secondary cycle of infection compared to the wild-type virus (Cory et al., 1994).

It is reported that speed of virus kill further increases in the presence of low doses of pyrethroid insecticides. Pyrethroid resistant insects are also killed more quickly by this combination. Since AaIT has a similar mode of action to the pyrethroid insecticides, the possibility of cross-resistance to the toxin was investigated. Instead of being susceptible to cross-resistance, AcNPV.AaIT was found more effective against resistant larvae than pyrethroid-susceptible larvae.

Two other neurotoxin genes that have been widely incorporated into the AcNPV genome are LqhIT2, which encodes a venome component of the yellow scorpian (*Leirus quinquestriatus hebraeus)* and Tox34,which expresses TxP-I toxin derived from the straw itch mite (*Pyemotes tritici*). Unlike AaIT, which is excitatory, LqhIT2 is a depressant-type toxin resulting in flaccid paralysis of the virus-infected insect. The straw itch mite uses a toxin to immobilize insects up to 1,50,000 times in size by causing muscle contraction and paralysis. The AcNPVs carrying either LqhIT2 or TxP-I have been evaluated against lepidopteran larvae, with each found to induce significantly enhanced speed of mortality (lethal time 30-40% less) than wild-type AcNPV. Thus, this engineered baculovirus not only hastened the demise of the infected insect larvae, it also significantly decreased the ability of insects to damage plants. These successes with toxins of relatively limited activity on lepidopterous larvae suggest that insect-selective peptides of far greater activity will prove useful with currently used expression systems.

Due to limited, effective host range of recombinant AcNPV, its potential to provide acceptable pest control might have utility against only two most AcNPV-permissive pest species, *T. ni* and *H. virescens*. One way to design a recombinant baculovirus with a more desirable target profile is to insert a toxin gene into a wild-type virus, which naturally exhibits high levels of pathogenicity to the pest complex one needs to control. For example HzNPV recombinants were recently engineered to carry either the LqhIT2 or AaIT toxin gene. These recombinants are potentially more effective biopesticide than AcNPV. AaIT for control of heliothine complex in cotton (DuPont, 1997; American Cyanamid, 1997)

3.2   Recombinant AcNPV Expressing Insect Hormones

Expression of a component of the insect endocrine system in recombinant baculoviruses is rationalized on a premise that in the event of a single component is overproduced, homeostasis will be disrupted, causing a deleterious effect on the insect. To test this possibility, recombinant AcNPVs were produced expressing eclosing hormone (which is involved in ecdysis and shedding of the cuticle during molting) and the prothoracicotropic hormone (PTTH) from *B. mori* (which stimulates production of ecdysone), respectively. However, none of these recombinant baculoviruses showed much enhancement in efficacy as compared to wild-type virus.

3.3 Deletion of egt Gene from AcNPV Genome

Insecticidal activity of the baculovirus can also be increased by the deletion of a gene(s) from the viral genome that alters the normal life cycle of the insect larva, but is not essential for virus replication.

The baculovirus AcNPV produces an enzyme ecdysteroid UDP-glucosyl-transferase (egt), which inhibits or delays molting and pupation in the host larvae. Expression of this protein prevents the host insect from molting and effectively keeps the insect in the feeding state. The deletion of the egt-gene from the viral genome allows the infected larvae to begin molting, which results in feeding cessation i.e. an increase in insecticidal activity, as compared to wild-type AcNPV-infected insects, which continue to feed throughout the period of infection. It was demonstrated that deletion of the *egt*-gene of AcNPV caused infected fall armyworm (*Spodoptera frugiperda*), a 10-20% reduction in lethal time, a 40% reduction in feeding damage and a 30% reduction in progeny viruses produced (O'Reilly and Miller, 1991).

In a field trial, Treacy et al. observed that while egt-deleted AcNPV provided more consistent control of *T. ni* on lettuce and cabbage, differences in efficacy between the wild-type and egt-deleted AcNPV were marginal. They concluded that any further changes that might be made in the baculovirus genome (e.g., gene insertion), to enhance killing speed beyond that seen with egt-deletion, should produce a more commercially acceptable biopesticide (Treacy, 1999).

American Cyanamid carried out field trials using cotton, tobacco, lettuce and cabbage as crops. A cotton trial in 1996, showed that treatment with recombinant virus egt-minus AcNPV containing the AaIT alone produced control of *Heliothis virescens* which was equivalent to that in a chemically treated control plot with acephate (American Cyanamid, 1997).

The concept of gene-deletion construct is more easily acceptable than for example a gene-addition construct, by the regulatory agencies. The egt-deleted recombinant AcNPV was the first genetically modified baculovirus to be field tested in the United States. Thus, it is likely to be the first recombinant baculovirus approved for commercial use, as a bio-pesticide.

3.4   Recombinant AcNPV Expressing Juvenile Hormone Esterase

Insects undergo hormonically controlled metamorphosis during development. This aspect of insect biology has been exploited for the development of faster-acting recombinant baculoviruses.

Juvenile hormone (JH) is intrinsically involved in regulation of gene expression in both larval and adult insects. From late embryonic life through the final larval instar, juvenile hormone is produced and secreted and maintained at measurable levels, resulting in the retention of larval characteristics. The reduction in JH titer is a key event in insect development that leads ultimately to termination of feeding and metamorphosis from larval to pupal stage. Juvenile hormone esterase (JHE) increases as JH titer decreases. The expression of JHE in a baculovirus vector, is to function as an anti-JH agent.

The coding gene for JHE from the tobacco budworm (*Heliothis virescens*) was expressed in various baculovirus constructs, including AcNPV. Despite high levels of expression of JHE, no significant improvement of the insecticidal efficacy was observed for neonate larvae of *T. ni.* or other insect species tested.

3.5   Environment Risk Assessment of Recombinant Baculoviruses

The potential hazards associated with chemical and viral pesticides differ mainly in the fact that viral pesticides replicate and, therefore, environmental contamination can increase in concentration and area. A serious concern is whether or not a recombinant protein, such as a toxin produced in the virus-infected insect, will present a hazard to other species in the environment by persistence. This carries an additional concern because once established in the environment, there is a finite probability that it cannot be removed if that is desired. In order to assess the risk of the release of a genetically modified organism, genetic modifications need to be evaluated in relation to two key ecological and environmental issues:

(a)   In what respect do the modified and wild-type baculoviruses differ in their effects on susceptible species and populations? and

(b)   Has the genetic modification altered the capacity of the baculovirus to persist and spread in the environment

Considerable research has been conducted to determine both absolute and relative impact of AcNPV-AaIT with respect to wild-type AcNPV, on survival of non-lepidopteran invertebrates. All the studies reported indicate that atleast this particular recombinant baculovirus exhibits the same host range as its wild-type counterpart. The range of virulence exhibited by AcNPV-AaIT, even in Lepidoptera, appears to be unaltered as compared to wild-type AcNPV. It has been shown that dosage-mortality responses were similar for AcNPV-AaIT and wild type AcNPV (Treacy, 1999).

Fuxa et al. (1998) tested the capability of recombinant baculoviruses to compete with a wild-type baculovirus for a host-insect resource in a green house microcosm. Two recombinant baculoviruses, AcNPV expressing a scorpian toxin (AcNPV.AaIT) and AcNPV expressing the mutant form of an insect derived juvenile hormone esterase (AcNPV.JHEKK) were tested with a wild-type *Autographa californica* nucleopolyhedrovirus (AcNPV.WT) on cabbage looper (*Trichoplusia ni*) larvae. They found that AcNPV.WT out-competed the recombinant viruses for a niche in a green house microcosm. This finding supports that the foreign genes such as AaIT is of negative value to the virus and probably ensures a competitive disadvantage of recombinant baculoviruses persisting in the agro-eco system relative to wild-type AcNPV.

4   Concluding Remarks

Recombinant baculoviruses represent a valuable technology that has great potential for effective integration into pest management system. Both the recombinant and wild-type viruses are more environmentally attractive than chemical pesticides and may represent a major step forward towards a more sustainable agriculture. Much of the research directed at improving pesticidal properties of baculoviruses has been conducted with AcNPV. However, recombinant AcNPV may have limited effective host range and utility against only pest species such as *T. ni* and *H. virescens*. This would present a barrier to its commercialization. A recombinant baculovirus with more desirable target profile would be the one, which exhibits high levels of pathogenicity to the pest complex such as for the control of the heliothine complex in cotton. For example, HzNPV recombinants carrying either LqhIT2 or AaIT toxin-gene have potential to be more effective biopesticides for pest control in cotton.

Commercialization potential of recombinant baculoviruses technology at the current level could be attained by its deployment in rotational-use strategies in control of lepidopteran pest species, for relaxing the selection pressure exerted by synthetic pesticides or even *Bt* on the pest populations. These could also be utilized in binary mixtures with certain chemical insecticides. Augmentation of transgenic crop varieties and establishment of binary transgenic host-plant resistance/biocontrol system could provide another use for them.

Optimization of viral shelf life, better suspension of the virus and increased stability under field conditions are some of the other challenges that would require satisfactory resolution by formulation technologists.

References

1. American Cyanamid, 1997. "Notification to conduct small-scale testing of a genetically engineered microbial pesticide, in *Federal Register*, United States Environment Protection Agency, Washington, DC, 62, 39,518.
2. Bonning, B. C. and B. D. Hammock, 1992. "Development and potential of genetically engineered viral insecticides", *Biotechnology and Genetic Engineering Reviews*, 10, 455-489.
3. Bonning, B. C. and B. D. Hammock, 1996. "Development of recombinant baculoviruses for Insect Control", *Annu. Revs. Entomol.* 41, 191-210.
4. Chakraborty, S., A. Kanhaisingh, P. I. Greenfield, S. Reid, C. Monsour and R. Teakle, 1995. "*In vitro* production of wild-type Heliothis baculoviruses for use as biopesticides", Australasian *Biotechnol.*, 5(2), 82-86.
5. Cory, J. S., M. L. Hirst, T. Williams, R. S. Hails, D. Goulson, B. M. Green, T. M. Carty, R. D. Possee, P. J. Cayley and D. H. L. Bishop, 1994. "Field trial of a genetically improved baculovirus insecticide", *Nature*, 370, 138-140.
6. Crook, N. E. and P. Jarret, 1991. "Viral and bacterial pathogens of insects", *J. Applied Bacteriology, Symposium Supplement,* 70, 91S-96S.
7. Cunningham, J. C., 1995. "Baculoviruses as microbial insecticides", in *Novel Approaches to Integrated Pest Management*, R. Reuveni, ed., CRC Press, Boca Raton, pp. 261-292.
8. DuPont Agricultural Products, 1997. "Notification to conduct small-scale field testing of genetically engineered microbial pesticides", in *Federal Register*, United States Environment Protection Agency, Washington, DC, 62, 23,448-23,449.
9. Federici, B., (1999). "Naturally occurring baculoviruses for insect pest control", in *Biopesticides: Use and Delivery*, F. R. Hall and J. J. Menn, eds., Humana Press, Totowa, NJ, pp. 301-320.
10. Friedlander, B. P., Jr., 1997. "New biological control of insect pests is discovered – Enhancin protein attacks the immune system; virus kills the insect", *Cornell University Science News*, http://www.news.cornell.edu/Science /June97/Enhancin Granados.bpf.html
11. Fuxa, J. H., S. A. Alaniz, A. R. Richter, L. M. Reilly and B. D. Hammock, 1998. "Capability of recombinant insect viruses for environmental persistence/transport", http://nbiap.biochem.vt.edu/brarg/brasym96/fuxa96 htm.
12. Gallo, L. G., B. G. Cossaro, P. R. Hughes and R. R. Granados, 1991. "In vivo enhancement of baculovirus infection by the viral enhancing factor of a granulosis virus of the cabbage looper, *Trichoplusia ni* (Lepidoptera: Noctuidae)", *J. Invertibr. Pathol.*, 58, 203-210.
13. Georgis, R., 1996. "Present and future prospects of biological insecticides", *Conference in Biological Control*, Cornell community, April, 11-13, 1996, http://www.nysaes.cornell.edu/ent/ bcconf/talks /georgis.html.
14. Hammer, D. A., T. J. Wickham, M. L. Shuler, H. A. Wood and R. R. Granados, 1995. "Fundamentals of baculovirus-insect cell attachment and infection", *in Baculovirus Expression Systems and Biopesticides*, M. L. Shuler, ed., Wiley-Liss, New York, NY, pp. 103-119,

15. Hoover, K., R. Herrmann, H. Moskowitz, B. C. Bonning, S. S. Duffey, B. F. McCutchen and B. D. Hammock, 1996. "The potential of recombinant baculoviruses as enhanced bioinsecticides", *Pestic. Outlook*, June 1996, 21-26.

16. Hostetter, D. L. and P. Putler, 1991. "A new broad-spectrum nucleo polyhedrosis virus isolated from a celery looper *Anagrapha falcifera* (Kirby) (Lepidoptera: Noctuidae)", *Environ. Entomol.*, 20, 1480-1488.

17. Jacob, T. K., 1996. "Introduction and establishment of baculoviruses for the control of rhinoceros beetle, *Oryctes rhinoceros* (Coleoptera: Scarabaeidae) in the Andaman Islands (India)", *Bull. Entomol. Res.*, 86, 257-262.

18. Maeda, S., 1995. "Further development of recombinant baculovirus insecticides", *Curr. Opin. Biotechnol.*, 6(3), 313-319.

19. Medugno, C. C., J. M. G. Ferraz, A. de H. N. Maia and C. C. L. Freitas, 1997. "Evaluation of a wettable powder formulation of the nuclear polyhedrosis virus of *Anticarsia gemmatalis* (Lep.: Noctuidae), *Pestic. Sci.*, 31, 153-156.

20. Menn, J. J., 1996. "Biopesticides: has their time come?", *J. Environ. Sci. Health*, B 31(3), 383-389.

21. Moscardi, F., 1999. "Assessment of the application of baculoviruses for control of Lepidoptera", *Annu. Rev. Entomol.*, 44, 257-289.

22. O'Reilly, D. R., and L. K. Miller, 1991. "Improvement of a baculovirus pesticide by deletion of the *egt* gene", *Bio/Technology*, 9, 1086-1089.

23. Park, E. J., C-M Yin and J. P. Burand, 1996. "Baculovirus replication alter hormone-regulated host development", *J. Gen. Virol.* 77(3), 547-554.

24. Pingel, R. L., and L. C. Lewis, 1997. "Field application of *Bacillus thuringiensis* and *Anagrapha falcifera* multiple nucleopolyhedro virus against the corn earworm (Lepidoptera: Noctuidae)", *J. Econ. Entomol.*, 90(5), 1195-1199.

25. Possee, R. D., A. L. Barnett, R. E. Hawtin and L. A. King, 1997. "Engineered baculoviruses for pest control", *Pestic. Sci.*, 31, 462-479.

26. Shapiro, M., 1995. "Radiation protection and activity enhancement of viruses", in *Biorational Pest Control Agents: Formulation and Delivery*, F. R. Hall and J. W. Barry, eds., American Chemical Society, Washington, DC, pp. 153-164

27. Treacy, M. E., 1999. "Recombinant baculoviruses", in *Biopesticides: Use and Delivery*, F. R. Hall and J. J. Menn, eds., Humana Press, Totowa, N.J., pp. 321-340.

28. Vail, P. V., D. F. Hoffmann and J. S. Tebbets, 1996. "Effects of a fluorescent brightner on the activity of *Anagrapha falcifera* (Lepidoptera: Noctuidae) nuclear polyhedrosis virus to four noctuid pests", *Biol. Control*, 7(1), 121-125.

29. Van Beek, N. A. M., and P. R. Hughes, 1998. "The response time of insect larvae infected with recombinant baculoviruses", *J. Invertebr. Pathol.*, 72, 338-347.

30. Volkman, L. E., 1997. "Nucleopolyhedrovirus interactions with their insect hosts", *Adv. Virus Res.*, 48, 313-348.

31. Wang, P. and R. R. Granados, 1997. "An intestinal mucin is the target substrate for a baculovirus enhancin", *Proc. Nat. Acad. Sciences*, 94(13), 6977-6982.

32. Washburn, J. O., B. A. Kirkpatrick, E. Haas-Stapleton and L. E. Volkman, 1998. "Evidence that stilbene-derived optical brightener, M2R, enhances *Autographa californica* M Nucleopolyhedrovirus infection of *Trichoplusia ni*

and *Heliothis virescens* by preventing sloughing of infected midgut epithelical cells, *Biol. Control*, 11 (1), 58-69..

33.   Weeden, Sheltn and Hoffmann, 1998. "Baculoviruses, Biological control. A guide to natural enemies in North America", Cornell Univ., http://www.nysaes.cornell.edu/ent/biocontrol/pathogens/baculoviruses.html.

34.   Wood, H. A., 1995. "Development and testing of genetically improved baculovirus insecticides", in *Baculovirus Expression Systems and Biopesticides*, M. L. Shuler, ed., Wiley-Liss, New York, pp. 91-102.

35.   Wood, H. A. and P. R. Huges, 1995. "Development of novel delivery strategies for use with genetically enhanced baculovirus pesticides", in *Biorational Pest Control Agents: Formulation and Delivery*, F. R. Hall and J. W. Barry, Eds., American Chemical Society, Washington, DC.

7

Biofungicides

1   Introduction

Increasing instances of resistance development and environmental concern are the major driving forces behind the developments of biofungicides for the control of plant diseases. Several scientific groups searching for biological means for control have addressed a number of significant fungal diseases. An overview of the progress achieved in commercializing biofungicides and control of some economically important plant diseases is provided in this chapter.

Diseases of the plants are the result of the interaction of a pathogen, a susceptible host plant and the environment. In control of entities, biocontrol agents are used to target both the disease (a process) and the organism (pathogen). However strategies for controlling the disease process (therapy) can differ from that used to control the pathogens (C. L. Wilson, 1997). The plant pathogens cause problems only under certain environmental conditions although most require free water in the environment to become active (Figure 7.1).

All living organisms are subject to disease, parasites and predators. Each of these natural means of control can be exploited to protect crops from pests. Microbial antagonism and hyperparasitism, phenomenon that exploit the effects of one or more organisms in the environment against the pathogen, comprise a viable means for control of plant diseases. Organisms that parasitize

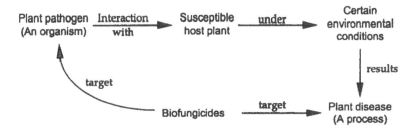

Figure 7.1   Biofungicides target both plant disease and plant pathogen.

and destroy the pathogen, are applied before planting the crop. On the other hand, organisms that function as antagonist by competing with the pathogen for supply of nutrients or space; or inhibiting its growth by antibiosis where antagonists secrete metabolites harmful to pathogens, are applied at the time of planting. Antagonistic microorganisms that compete or directly attack the pathogen are mixed with the soil, added to furrows, used as seed treatment and sprayed into foliage or fruit. A number of reviews on biological control of plant pathogens and plant diseases have appeared in the literature (Be'langer et al., 1998, Chalutz and Droby, 1998; Backman et al., 1997; El-Ghauth, 1997; Mehrotra et al., 1997; Whipps, 1997; Utkhede, 1996; Malony, 1995 and Cook, 1993).

     Like other disease management methods, biocontrol ideally, aims to suppress disease sufficiently so that avoidable yield losses are minimized and crop quality is maintained at an acceptable level. Biological systems however, tend to be unreliable due to the effect of widely varying environmental influences on their efficacy. For this reason, biofungicides have been more successful with protected crops and post-harvest infestation control, where there is a degree of control on environmental parameters. Biofungicides complement synthetic fungicides for the control of disease on agronomic and horticultural crops. Therefore, they are preferred for integrated use in combination with fungicides (Figure 7.2).

     Commercial biofungicides are characterized for their consistent suppression of fungal pathogens under field conditions, and easy mass production in standard fermentation facilities. These can be classified into three categories based on their target applications: soilborne pathogens, foliar diseases and post-harvest rots during storage. All the three categories have been extensively studied and are reviewed in the following paragraphs. Some of the important soil-borne and post-harvest diseases have also been discussed. A discussion on two important biocontrol agents, one a fungal and another a

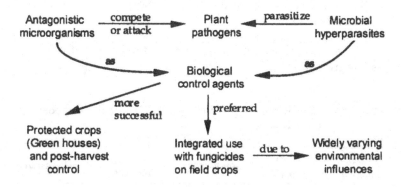

Figure 7.2   Biological control is more reliable in protected crops.

bacterial system has been included. The production and formulation aspects of bacterial and fungal biofungicides have required considerable developmental efforts and these aspects have also been reviewed here.

2   Biological Control of Soil-Borne Diseases

In the field, many ineradicable soil-borne diseases occur such as those caused by fungal plant pathogen species of *Fusarium, Pythium, Verticillium, Rhizoctonia, Macrophomina, Phytophthora* and others, which cause either seed rot before germination or seedling rot after germination. The diseases are often termed pre- and post-emergence damping-off, or seedling blights. The soil-borne plant pathogens cause major economic losses in agriculture. The present methods for control of these diseases are inadequate and also cause non-target side effects on beneficial microorganisms and the environment. Thus, there is a need for a suitable alternative to synthetic pesticides. Alternative disease control is sometimes possible through development of crop plants that are resistant to disease. Unfortunately, resistant plants are not available for many diseases caused by soil-borne plant pathogens.

Biological control is another means of controlling soil-borne fungal disease, which is presently receiving much attention. Biological control has been more successful in the plant rhizosphere employing both bacterial and fungal biocontrol agents, than in the phyllosphere.

2.1   Bacterial and Actinomycetous Biofungicides
      for Soil-borne Pests

A list of commercial bacterial biofungicides for control of soil-borne plant
diseases is given in Table 7.1.

Table 7.1   Commercial bacterial biofungicides for control of soil-borne
plant diseases.

Microbial Antagonist	Trade Name	Manufacturer	Target Pathogen	Diseases/Crops
Bacillus subtilis (Strain GBO3)	Kodiak	Gustafson, Inc., Plano, TX, USA	*Pythium ultimum*, *Rhizoctonia solani* and *Fusarium* spp	Seedling diseases and late season *Rhizoctonia* root rot in cotton
B. subtilis (Strain GBO7)	Epic			Seedling diseases of cotton and legumes
Bacillus subtilis (Strain QST713)	Serenade	AgraQuest, Davis, CA, USA		Powdery and downy mildew, late and early blight and brown rot
Agrobacterium radiobacter (Strain K 84)	Norback 84-C	New Bio Products, Corvallis, OR, USA	*Agrobacterium tumefaciens*	Preventive (not curative) control of crown gall in grapes, blueberries, bushberries, deciduous trees and ornamental plants
	Galltrol	AgBioChem, Orinda, CA, USA		
A. radiobacter (Strain K 1026)	NoGall	Bio-Care Technology, Somersby, Australia	*A. tumefaciens*	Also prevents transfer of agrocin resistance to *A. tumafaciens*

Microbial Antagonist	Trade Name	Manufacturer	Target Pathogen	Diseases/Crops
Enterobacter aerogenes (Strain B 8)			*Phytophthora cactorum*	Crown and root rot of fruit trees (e.g. apple) by soil and trunk drench
Pseudomonas fluorescens	Dagger G	Ecogen, Langhorne PA, USA	*Pythium ultimum, R.. solani*	Seedling diseases of cotton
P. fluorescens NCIB	Conquer	Mauri Foods, North Ryde, Australia	*Pseudomonas tolaasil*	Bacterial blotch of mushroom
P. fluorescens NCIB 12089	Victus	Sylvan Spawn, Kittanning, PA	*P. tolaasil*	Bacterial blotch of mushroom
P. fluorescens WCS374	BioCoat	S & G Seeds, The Netherlands	*Fusarium oxysporum f.sp. raphani.*	*Fusarium* wilt in radish and carnation
Burkholderia cepacia (Wisconsin M36)	Deny (formerly Blue Circle)	Stine Microbials, Haskins, KS, USA	*Fusarium* spp., *Pythium* spp. and *Rhizoctonia solani*	Disease caused lesion, spiral, lance and sting nematodes
B. cepacia (Strain 679-2)	Leone	Dominion BioSciences, Walnut Creek, CA, USA	*Botrytis, Septoria* and *Phytophthora,* spp.	*Botrytis,* and *Phytophthora* on potatoes and *Septoria* on wheat
Streptomyces griseoviridis K-61	Mycostop	Kemira Agro Oy, Helsinki, Finland AgBio development, Westminster, CO, USA	*Fusarium oxysporum* f. sp. *dianthi* and *Alternaria brassicicola* and *Phytophthora, Thielaviopsis* and *Macrophomina* spp.	Wilt of carnation; root diseases and foot rot of cucumber, *Fusarium* disease in wheat and in tropical plantation-crops; damping off diseases in cauliflower, broccoli, cabbage and ornamental plants.

Brannen and Kenney (1997) of Gustafson have reported successful use of Kodiak®, a *Bacillus subtilis* strain GBO 3, as a bacterial seed treatment for control of seedling diseases of cotton, beans and other crops. Kodiak ® is reported to be effective against pathogenic species of *Rhizoctonia* and *Fusarium*. The product is formulated as wettable powder and is compatible with several seed treatment fungicides. Similarly, *Bacillus subtilis* strain A13 is inhibitory *in vitro* to several plant pathogens and is being marketed under the name Quantum 4000® for peanuts and Epic® for seedling diseases in cotton for the control of *Fusarium* spp. and *Rhizoctonia* spp. (Utkhede, 1996). A *B. subtilis* strain QST713 has been found to be effective in controlling various foliar plant pathogens, such as *Botrytis*, powdery mildews, fire blight and some downy mildews. AgraQuest is marketing this strain as Serenade®. Another biofungicide based on a *B. pumilus* strain has been in field trials and is expected to be available in the market in 2000 under the trade name Sonata (Marrone, 1999). *B. cereus* UW85 is another gram-positive bacterium that has been found effective in controlling *Phytophthora* damping off and root rot of soybeans under diverse field conditions (Emmert and Handelsman, 1999). Presently, the greatest commercial usage of rhizosphere-colonizing bacterial biocontrol agents, largely *Bacillus* spp. are in combination with seed dressing chemical fungicides. The fungicides are still necessary because *Bacillus* spp. are too slow in activating and effectively colonizing the seed and emerging root to deal with fast-growing pathogens. Dicarboximides such as vinclozolin and iprodion are best suited for use along with *Bacillus* biocontrol agents (Seddon et al., 1995). The marketing strategy has been to address season-long root health, reducing fungal pathogen damage caused after the seed-treatment fungicides have dissipated, usually occurring more than two weeks after planting.

Agrobacterium radiobacter strain K 84 is another example of economic biological control with a bacterial antagonist. The commercial preparation Galltrol® or Norback 84-C® is reported to be effective in controlling crown gall (*A. tumefaciens*) on stone fruits, commonly attributed to their production of agrocin, an antobiotic. A recombinant *A. radiobacter* (derivative K 1026) was engineered to prevent transfer of agrocin resistance, from the bacteria *A. radiobacter* to *A. tumefaciens* (Jones et al., 1989; Maloney, 1995). The engineered strain was marketed in Australia in 1988 under the trade name NoGall®. Yet another bacterial antagonist *Enterobacter aerogenes* strain B8 reportedly controlled crown and root rot of fruit trees caused by *Phytophthora cactorum* (Utkhede, 1996). When *E. aerogenes* (B8) was alternated with metalaxyl, *P. cactorum* infection was significantly reduced and fruits yield of apple trees increased.

Ecogen's Dagger G®, a *Pseudomonas fluorescens* granular peat-based formulation controls seedling pathogens of cotton. Similarly, another commercial preparation of *P. fluorescens*, Conquer® has been found effective for control of *Pseudomonas tolassil* on mushrooms in Australia. A *P. fluorescens* WCS 374 preparation BioCoat® is applied as a seed coating to radish and carnation for control of *Fusarium*.

Burkholderia cepacia, a bacterium formerly known as *Pseudomonas cepacia* was renamed in honor of Walter Burkholder, a Cornell University plant pathologist, who first reported it to be the cause of sour skin of onions. It has been shown to be an effective biocontrol agent against soil-borne as well as foliar and post-harvest diseases. Many strains of *B. cepacia* produce one or more antibiotics active against a broad range of plant pathogenic fungi. Deny®.is registered with U.S. EPA for use as microbial biofungicide.

Kemira Agro Oy of Finland has put in the market a freeze-dried actinomycete product Mycostop® based on *Streptomyces griseoviridis*. It is a wettable powder containing mycelium and spores and can be applied to seed as a dry powder or suspended in water and used as a dip, spray or drench. It is effective in controlling wilt diseases of carnation, root and foot rot of cucumber, fusarial diseases in wheat and in tropical plantation crops, damping-off diseases in cauliflower, broccoli, cabbage and ornamental plants.

2.2 Fungal biofungicides for soil-borne pests

The fungal organisms researched as biocontrol agents are primarily filamentous fungi, such as, *Gliocladium virens* and *Trichoderma harzianum*. However, there are also examples of some yeast-like fungi such as *Ampelomyces quisqualis*. The environmental conditions such as temperature and moisture affect the growth and survival of fungal biofungicides and limit their efficacy. The application of these biofungicides is for control of root-infecting pathogens, e.g.. *Pythium, Rhizoctonia* and foliar fungal pathogens, e.g. powdery mildew and *Botrytis*. Various mycofungicides commercialized for control of soil-borne plant diseases are tabulated in Table 7.2.

Cornell University researchers have produced a hybrid strain of a filamentous fungus, *Trichoderma harzianum 1295-22*, by protoplast fusion between *T. harzianum* strains T-12 and T-95. It has efficacy against a wide range of plant pathogenic fungi including, *B. cinerea*, protects the root system against *Fusarium, Pythium* and *Rhizoctonia* on a number of crops including corn, soybeans, potatoes, tomatoes, beans, cotton, peanuts, turf, trees, shrubs and other transplants and ornamental crops. It is compatible with many standard chemical seed treatments (e.g. Captan) and is applied directly over chemically treated seeds at the time of planting (Sivan et al., 1991). Two commercial products RootShield® and Bio-Trek 22G® are registered by BioWorks, Inc., Geneva, NY. There is another product being developed by Mycotech Corporation is a blend of two complementary strains of *Trichoderma* under the trade name EcoGard™. It has been found effective against various soil borne

Table 7.2   Commercial mycofungicides for control of soil-borne plant diseases

Fungal Pathogen	Trade Name	Manufacturer	Target Pathogen	Disease/Crops
Fusarium oxysporum (Nonpathogenic strain Fo47)	Fusaclean	Natural Plant Protection, Nogueres, France	*F. oxysporum*	*Fusarium* wilt of vegetable and flower crops in green house
Gliocladium virens GL21	SoilGard	Thermo Trilogy, Columbia, MD	*Pythium* spp., *Rhizoctonia solani* and *Sclerotium rolfsi*	Damping off and root rot of ornamental and food crop plants in green house.
Trichoderma harzianum (Strain 1295-22)	RootShield and T-22 Planter Box	BioWorks,. Geneva, NY, USA	*Pythium* spp., *Rhizoctonoa solani* and *Sclerotinia homoeocarpa*	Damping off and root rot of corn, bean and vegetables
Trichoderma harzianum and T. viride	Promote	JH Biotech, Ventura, CA, USA	*Pythium, Rhizoctonia* and *Fusarium*	Greenhouse crops, nursery transplants and seedlings
T. harzianum (ATCC 20476) and *T. polysporum* (ATCC 20475	Binab T	Bio-Innovation AB, Algaras, Sweden	*Chondrostereum purpureum, Endothia parasitica, Verticillium malthousei, Heterobasidion annosum, Armillaria mellea and Ceratocystis ulmi.*	Tree wood decays, silver leaf of plum and apple; Chestnut blight, pinewood root decay
Pythium oligandrum	Poly-gandron	Vyskumny ustav, Piestany, Slovak	*P. ultimum*	Damping-off of sugar beet

Fungal Pathogen	Trade Name	Manufacturer	Target Pathogen	Disease/Crops
Peniophora (Phlebia) gigantea	RotStop	Kemira Agro Helsinki, Finland	*Heterobasidion annosum*	Stem and root rot of pine
	P.g. suspension	Ecological Labs., Dover, UK		
Aspergillus niger AN27	Kalisena	IARI, New Delhi, India	*F. oxysporum, F. solani, Macrophomina phaseolina, Pythium* spp., *R. solani* and *S. sclerotiorum*	Cereals, millets, pulses, oil-seeds, fruits, tubers, vegetables and ornamentals

diseases and also gives good control of *Botrytis* on strawberries (Mycotech, 1999). Similarly, a *T. harzianum* product has been reported to be effective against *Sclerotium rolfsii* in chick pea (Mukhopadhyay et al., 1994).

SoilGard® (Thermo Trilogy) developed by Biocontrol of Plant Diseases Laboratory (BPDL) of USDA and Grace Biopesticides, is a granular formulation for control of soilborne plant pathogens, including *Pythium spp., Rhizoctonia solani, Sclerotium rolfsii* and others by the common soil saprophytic fungus, *Gliocladium virens* GL21. It produces an antibiotic gliotoxin that is implicated as the key factor in its biocontrol activity against *Pythium ultimum* and *Rhizoctonia solani* (Wilhite et al., 1994). *Phlebia gigantea* was one of the first microbes used for control of *Heterobasidion annosum* in trees. Since the pathogen can colonize freshly cut stumps and then spread through root grafts, the product RotStop® is applied to freshly cut stumps.

A few Pythium species are known to be aggressive mycoparasites. *Pythium nun,* attacks *Pythium ultimum* and *P. vexans*, by coiling around the hyphae of the host, forming appressoria-like structures and then parasitizing the hyphae (Lifshitz et al., 1984). Similarly It is reported that sugarbeet seeds coated with oospores of *P. oligandrum* effectively controlled preemergence damping-off caused by *P. ultimum* (Martin and Hancock, 1987). *Pythium oligandrum* is an aggressive mycoparasie and can be used as a biological control agent of several plant pathogenic fungi. A videomicroscopic study depicts how Pythium oligandrum recognises its host fungi and destroys them (Figure 7.3) (Laing and Deacon, 1991). A product based on *P. oligandrum* has been commercialized in the trade name Polygandron® by Vyskumny ustav of Slovak Republic.

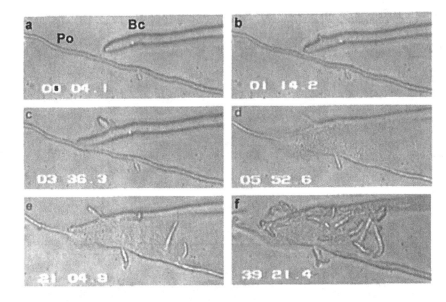

Figure 7.3   Videotaped interaction between *Pythium oligandrum* (Po) and the plant
pathogenic fungus *Botrytis cinerea* (Bc) on a thin film of water agar.
(Curtsey: J. W. Deacon). (a and b) The hyphal tip of Bc contacted the
hypha of Po, (c) The tip of Bc suddenly lysed (d and e) expelling the
protoplasm, (f) Po produced many hyphal branches using the spilled
protoplasm as a nutrient source

Kalisena®, a product based on *Aspergillus niger* isolate AN27 is
reported to control several soil borne pathogens including *F. oxysporium*, *F.
solani*, *Pythium spp.*, *R. solani* and *S. sclerotiorum* in diverse group of crop
plants (Sen et al., 1996). The product is reprted to be in early stages of
commercialization.

3   Soil-borne Diseases

Some of the economically devastating soil borne diseases, such as *Fusarium* wilt,
Phytopthora, *Sclerotinia* and *Pythium* diseases and their control are discussed in
following paragraphs.

3.1 Fusarium wilts

Fusarium wilt and fusarium root-rots are among the most severe plant diseases caused by pathogenic strains of *Fusarium oxysporum*, one of the most common species among soil fungi in cultivated soil. These fungi have a large diversity of strains, all of them being saprophytic; many of them are parasitic and some of them pathogenic and induce either root rot or tracheomycosis. *Fusarium* prefers warm, dry conditions typical of summer. The wilt inducing *F. oxysporum* strains show a high degree of host-specificity; and certain saprophytic strains of the same have been tried for reducing the pathogenic population.

The concept of certain soils being 'conducive' or 'suppressive' to disease refers to the soils that favor the expression of disease, or to the soils that are inhospitable to some plant pathogens respectively. In suppresive soils, a pathogen cannot become established or if established cannot initiate disease. Presence of a large population of non-pathogenic *F. oxysporum* in soil, suppresses fusarium wilts. The modes of action include a competition between the pathogenic and non-pathogenic *F. oxysporum* for nutrients and infection sites in the rhizosphere and also induction of resistance in the plant (Figure 7.4).

Fluorescent *Pseudomonas* spp. are among the most abundant bacteria in the rhizosphere. They are characterized by the production of yellow-green pigments (pyoverdines). These fluoresce under UV light and function as siderophores. It has been established that some strains of fluorescent *Pseudomonas* spp. are able to inhibit the growth or activity of fungal pathogens including *F. oxysporum*. The main mechanism of the antagonistic expression by

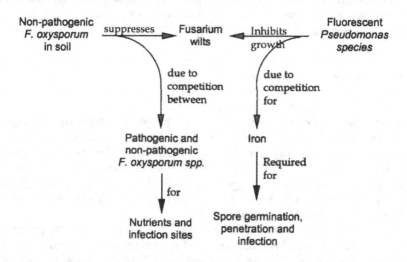

Figure 7.4   Competition mechanisms of suppression of *Fusarium* wilts

fluorescent *Pseudomonas* spp. against *F. oxysporum* is the competition for iron (Figure 7.4). It has been well established that fluorescent *Pseudomonas* spp. producing the catechol-hydroximate sidorophores are very efficient competitors for iron (Fe^{3+}) required for germination, penetration and infection (Baker and Griffin, 1995). Reducing availability of iron to pathogenic *F. oxysporum* results in an increased level of soil suppressiveness. Additionally, it is also likely that fluorescent *Pseudomonas* mediated induction of systemic resistance is also involved in the reduction of disease.

Lemanceau et. al. (1992,1993) demonstrated that a complementary effect of non-pathogenic *F. oxysporum* and fluorescent *Pseudomonas* spp. significantly contributed to the suppressiveness of soils to *Fusarium* wilt. They used a siderophore-deficient mutant of *P. putida* to demonstrate that competition for iron, resulting from the activity of the bacterial strain, controlled the efficacy of competition for carbon between strains of *F. oxysporum.*

The efficacy of biological control depends on the population density of the antagonistic microorganisms, but in case of *F. oxysporum* the ratio of population density (> 10) of the non-pathogenic versus the pathogen is more important than the absolute value of the population density itself (Alabouvette et al., 1998). Specific isolates of non-pathogenic *F. oxysporum* have been reported to be highly effective at controlling *Fusarium* wilt of tomato (50-80% reduction of disease incidence) and other crops (BARC, 1999). These results indicate that non-pathogenic *Fusarium* spp. have potential for further development as biological control agents and can be a viable alternative to fumigation with methyl bromide for several crops. Fusaclean®, a commercially available microgranular product based on nonpathogenic *Fusarium oxysporum* controls pathogenic *F. oxysporum* and *F. moniliforme* on basil, carnation, cyclamen and tomato (Table 7.2) (Fravel et al., 1998).

3.2   Phytophthora Diseases

Fungi in the genus *Phytophthora* cause important diseases on many economically important crops worldwide. The late blight pathogen *P. infestans* causes serious losses of potato crops. *Phytophthora* root, crown, and collar rots are common and destructive diseases of fruit trees. Disease development is favored by cool (15 to 25°C), wet, humid conditions and is most severe in poorly drained sites. Significant damage occurs only when soils are extremely wet or saturated. *Phytophthora* grows through the root destroying the tissue, which is then unable to absorb water and nutrients.

Many bacteria, actinomycetes and fungi are antagonistic to *Phytophthora,* and undoubtedly such systems are operative in nature. However, the development of effective biological control of *Phytophthora* species has been fraught with difficulties because of their ability to produce several forms of inoculum (zoospores, sprangia, chlamydospores, oospores and mycelium) rapidly and repeatedly, their ability to penetrate and infect a host plant within a few hours, their propensity to exist in soil in depths that allow them to escape most antagonists and in some cases their wide host range (Erwin and Riberio, 1996).

Crown and root rot of fruit trees is primarily caused by *Phytophthora cactorum* (Utkhede, 1986). Several fungi show promise for biological control of this pathogen by parasitizing and lysing its propagules in soil. For instance, *Trichoderma* and *Gliocladium* species suppressed *P. cactorum* on apple seedlings in glass house tests. A commerical preparation Binab T®, containing spores of *Trichoderma* spp. inhibited formation of sporangia and caused lysis of *P. cactorum* mycelium *in vitro* (Table 7.2). Binab T®, reduced the size of lesions (54%-57%) on inoculated apple twigs as well as on apple trees sprayed with it (Figure 7.5) (Orlikowski and Schmidle, 1985).

Although *Trichoderma* species antagonise *Phytophthora* spp., isolates of this fungus also have been shown to stimulate the formation of oospores in normally self-sterile isolates of *Phytophthora*. Response of *T. viride* on *P. cinnamomi* is reported to be both lysis of hyphae and induction of oospores (Reeves, 1975). However, production of these resting stores may compromise the effectiveness of biological control.

There is much evidence to demonstrate that most *Phytophthora* spp. are relatively non-saprophytic and do not grow well in competition with other microorganisms in soil. Simply filling in the transplanting holes with non-infested soil controlled the seedling disease of papaya caused by *P. palmivora*. The practice exploited the inability of *P. palmivora* to compete with naturally occurring microorganisms in soil (Ko, 1982).

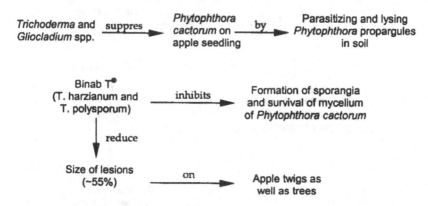

Figure 7.5   Mechanism of biological control of *Phytophthora* disease in apple seedlings.

3.3   *Sclerotinia Diseases*

As a group, *Sclerotinia* spp. are fungi, typically comprised of soil-borne pathogens that produce monocyclic diseases in the rhizosphere and phyllosphere. Diseases caused by *Sclerotinia* spp. are difficult to manage because of their widespread distribution, sporadic occurrence and long-term survival in soil. There are three economically important *Sclerotinia* species viz. *S. sclerotiorum, S. trifoliorum* and *S. minor*.

   Sclerotinia sclerotiorum is the most widespread and economically important pathogen within the genus *Sclerotinia* because of its widespread distribution and host range. It is considered an opportunistic invader of senescent or dead foliar and floral tissues. The epidemiology of disease is primarily associated with carpogenically germinated sclerotia that produce apothecia which release ascospores. Infection of *S. trifoliorum* also typically originates from ascospores, but germinating spores penetrate foliage directly and establish quiescent infections, eventually becoming more aggressive and developing into an expansive leaf and stem rot. On the other hand, *S. minor* is associated with eruptive or myceliogenic germination of sclerotia in the rhizosphere, which results in direct infection of roots and crowns of susceptible plants.

   The strategy adopted for biological control of diseases caused by *Sclerotinia* spp. is to reduce the concentration of initial inoculum by killing sclerotia or inhibiting their germination. Use of parasitic fungi or mycoparasites, have been found useful in reducing the number of sclerotia in infected soils (Figure 7.6).

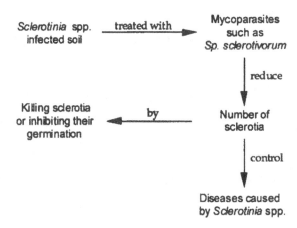

Figure 7.6   Mechanism of biological control of diseases caused by *Sclerotinia* spp.

One of the most effective mycoparasites is *Sporidesmium sclerotivorum* that behaves as an obligate parasite on sclerotia of *Sclerotinia* spp. After sporulating on the surface of the sclerotium, *Sp. sclerotivorum* can grow for upto 3 cm through soil to infect a new sclerotium. The ability of *Sp. sclerotivorum* to proliferate and grow through soil to infect new sclerotia, facilitates the epidemic destruction of sclerotia, regardless of whether the host plant is present or not (Fravel, 1998). In order to reduce the number of viable sclerotia in the subsequent crops, application of spore suspensions of *Sp. sclerotivorum* to the phyllosphere successfully controlled *Sclerotinia* diseases that do not involve ascospores in the disease cycle. This approach has been successfully used to control lettuce drop incited by *S. minor* (Zhou and Boland, 1998)

There are eight other fungi reported, that are antagonistic to *Sclerotinia* spp. These are *Teratosperma oligocladium*, *Coniothyrium minitans*, *Dictyosporium elegans*, *Penicillium citrinum*, *Talaromyces flavus*, *Gliocladium virens*, *Trichoderma* spp. and *Erwina herbicola*.

3.4   *Pythium Diseaes*

Species of *Pythium* are best known as pathogens of crop plants. They cause damping-off and other seedling diseases, and also progressively destroy the root tips of older plants, leading to 'decline' diseases of orchard trees and other perennial crops. It belongs to a group of fungi called the water molds. As the name implies, these fungi flourish in an environment in which water is constantly available. *Pythium* flourishes when soils are cool and wet, such as during the spring. It can infect seeds within 12 h. of planting. Sporangia of *Pythium ultimum* must germinate in order to incite seed and seedling rots. This germination occurs within 0.5-1 hr after exposure to exudates of germinating seeds and colonizes the seed very rapidly after planting (Harman and Nelson, 1994). In cotton and possibly other plant species, fatty acids are important germination stimulation molecules, with linoleic acid being the most stimulating. A biocontrol agent capable of inactivating the stimulatory activity of seed exudates and linoleic acid, would control damping off rot of seeds and seedling.

4   Biological Control of Aerial Plant Diseases

The availability of several relatively effective fungicides for use against a majority of foliar fungal pathogens, has been a disincentive to develop biocontrol agents for these pathogens. However, the now frequent occurrence of fungicide resistance, for example, to the benzimidazoles and dicarboximides, has necessitated the development of alternate strategies. Particularly, biocontrol of *Botrytis cinerea* that causes the disease grey mold and *Uncinula necator* and *Sphaerotheca fuliginea* causing pathogens of powdery mildews has been studied.

4.1   Grey mold disease

The necrotrophic *Botrytis cinerea* causes the disease grey mold which is a serious economic problem in a number of fruit crops, such as grapes, strawberry, raspberry, orchard fruits and green house crops such as tomatoes and potted plants. *Botrytis cinerea*, a deuteromycete fungus is an important plant pathogen with an exceptionally broad host range. Conidia of *B. cinerea*, the primary inoculum source, first attach to substrata by a hydophobic interaction that is easily disrupted and then, upon germination, attach strongly, through secretion of an extracellular matrix by germ tubes and appresseria but not the conidia themselves, known as fungal sheath. The enzymatic action occurs immediately after the pathogen comes into contact with the host and continues through the time of penetration. It typically requires exogenous nutrients during germination and germ tube elongation. These pathogens are subjected to competition for these nutrients with the indigenous saprophytic microbial community on foliar surfaces (M. Wilson, 1997).

Suppression of sporulation of *B. cinerea* has been effectively achieved through foliar applications of the saprophytic fungi *Trichoderma* spp., *Penicillium* spp. and *Gliocladium roseum* in various hosts, including strawberry, grapes and cucumber. *Trichoderma harzianum* isolate T-39 (Trichodex®) is commercially available for control of greymold of grapes (Table 7.3).

4.2   Powdery Mildews Disease

Biotrophic pathogens causing powdery mildews, rust and downy mildews are economically significant. Particularly, powdery mildew is a significant fungal disease affecting many crops. Special structures of the mildew penetrate the plant epidermal cells and feed on cellular tissue, causing dwarfing of the plant and the fruit, and cosmetic damage which is particularly undesirable on fresh produce. Plants severely infected with these pathogens have reduced yields. Powdery mildews such as *U. necator* and *S. fuliginea* are obligate parasites and typically do not require exogenous nutrients during germination. This precludes the use of nutrient competition as a biocontrol strategy, as used against *B. cinerea*. Instead, hyperparasitism as an approach has been successfully applied in biocontrol of powdery mildews on various plant hosts.

Table 7.3   Biocontrol agents of aerial diseases.

Microbial Antagonist	Trade Name	Manufacturer	Target Pathogen	Control
Pseudomonas fluorescens (Strain A 506)	Blight Ban A 506	Plant Health Technologies, Boise, ID, USA	*Erwinia amylovora*	Fire blight and frost injury in pear and apple
Ampelomyces quisqualis (Strain M 10)	AQ10	Ecogen, Langhorne, PA, U.S.A.	*Uncinula necator, Erysiphe cichora cearum*	Powdery mildew in apples, grapes and chillis
Trichoderma harzianum (Strain T 39)	Trichodex	Makhteshim Chemical Works, Beer Sheva, Israel	*Botrytis cinerea, Uncinula necator Erysiphe cichora-cinerea*	Greymolds/ powdery mildew in grapes, strawberries and cucumber
Mixture of *T. hamatum* + *Rhodotorula glutinis* + *B. megaterium*	Grey Gold	Eden Bioscience, Poulsbo, WA, U.S.A.	*B. cinerea*	

A fungal hyperparasite *Ampelomyces quisqualis* has been identified that infects most, if not all types of powdery mildews harmful to agricultural products. It infects and forms pycnidia (fruiting bodies) within powdery mildew hyphae. This parasitism reduces growth and may eventually kill the mildew colony. It reduces or eliminates applications of sulphur and DMI fungicides such as mycobutanil and fenarimol. Ecogen Inc. has commercialized an isolate M-10 of *A. quisqualis* (AQ10®), that has been found effective on grape powdery mildew caused by *U. necator* and is also reported to give good control of powdery mildew of cucumber at moderate disease pressure.

Verticillium lecanii, on the other hand is a polyphagous organism that attacks the structure of its hosts and parasitizes powdery mildew fungi and rusts. *V. lecanii* is reported to adequately control cucumber powdery mildew on a resistant cultivar, although requirement of high humidity remains a limiting factor in its efficacy (Verhaar et al., 1993).

Sporothrix flocculosa is a yeast like epiphytic fungus, that is a powerful antagonist of powdery mildews. It was found effective in controlling cucumber, rose and wheat powdery mildews. Temperature and humidity requirements are important constraints in maintaining the efficacy of *S. flocculosa*. It induced a rapid plasmolysis of powdery mildew pathogen cells, suggesting its mode of action by antibiosis. Some of the biocontrol agents commercialized for the control of aerial plant diseases are given in Table 7.3.

Effective biocontrol agents for control of serious foliar diseases such as grape downy mildew, potato late blight, wheat powdery mildew and apple scab remain to be identified (Froyed, 1996). However, *Fusarium proliforatum* was recently proposed for downy mildew control on grapes (Hofstein and Chapple, 1999).

5   Control of Post-Harvest Decay of Fruits and Vegetables

Fruits and vegetables are highly perishable while still maintaining active metabolism long after harvest. Post-harvest infection can occur either prior to harvest or during the harvesting and subsequent handling and storage. In general, most harvested commodities are resistant to fungal infection during their early post-harvest phase. However, during ripening and senescence, they become more susceptible to infection. Key infection sites for post-harvest diseases are wounds created by harvesting and handling, though some pathogens may directly penetrate. Post-harvest losses can be severe, if adequate handling and refrigerated controlled atmosphere storage facilities are lacking.

Several biological control approaches, including the use of antagonistic microorganisms, natural fungicides and induced resistance have been shown to have a potential as a antimicrobial preservative for harvested commodities. Development of antagonistic microorganisms has been the most studied, among these alternatives. Antagonistic microorganisms are selected for their ability to directly parasitize and compete with plant pathogens, as well as to induce resistance responses in the host to the disease. For enhanced biocontrol activity, antagonistic microorganisms are combined with natural fungicides such as chitosan, which has capability of 'turning on' host defenses to disease (C. L. Wilson, 1997).

The chances of success of biological control measures are high in the post-harvest environment. The partially controlled environment, particularly temperature and humidity may help shift the balance of the interactions between the host, the pathogen and the antagonist in favor of the antagonist. It is generally assumed that the mode of action of biocontrol of post-harvest diseases involve a complex mechanism, which comprises one or more of the following processes: antibiosis, nutrient competition, site exclusion, induced host resistance and direct interactions between the antagonist and the pathogen.

Many reports describe the inhibition of post-harvest diseases by antibiotic producing microorganisms, based on the *in vitro* test results, in which the antagonist inhibits the growth of the pathogen in culture. However, a debate has ensured, whether antibiotic-producing antagonists should be used on fruitsand vegetables, resulting in introduction of antibiotics into our food. This may result in resistance developments in humans to antibiotics as well as development of pathogen resistance (Kohl and Fokkema, 1998).

A number of antagonistic microorganisms have been identified and shown to be effective against various post-harvest pathogens on a variety of harvested commodities (Table 7.4). These include apple, apricot, avocado, banana, cabbage, carrot, celery, cherry, citrus, gerbera, grape, kiwi, litchi, mango, nectarine, peach, pear, pineapple, plum, potato, roses, strawberry, sweet potato, tomato, tulip bulbs and grains.

Utkhede (1996) has reviewed successful application of antagonistic bacterium *Pseudomonas syringae* (strain ESC11) for control of blue-mold rot (*Penicillium expansum*), on wounded pear fruit as well as on Empire apples and citrus fruits. *P. syringe* also provided control of greymold rot (*Botrytis cinerea*) on Bosc pears. EcoScience has commercialized this product under the trade name BioSave 11[®].

Table 7.4   Biocontrol agents of post-harvest diseases of fruits and vegetables

Microbial Antagonist	Trade Name	Producer	Target pathogen	Disease/ Produce
Pseudomonas syringae (ESC 11)	Bio-Save 110	EcoScience, Worcester, MA, U.S.A.	*Penicillium expansum, P. digitatum, Botrytis cinerea* and *Mucor piriformis*	Blue mold rot, green mold, grey mold and mucor rot of apples
P. syringae (ESC 10)	Bio-Save 1000	EcoScience, Worcester, MA, U.S.A.	*Penicillium digitatum, P. italicum* and *Geotrichum candidum*	Green mold, blue-mold and sour rot of citrus fruits
	Bio-Save 100			Blue mold and grey mold in pears
Burkholderia cepacia (Wisconsin Iso J82)	Deny	CCT Corp., Carlsbad, CA, U.S.A.	*Botrytis cinerea*	Grey mold rot of apples and pears
Candida oleophila I-182	Aspire	Ecogen, Langhorne, PA, U.S.A.	*Botrytis* spp and *Penicillium*. spp.	Post-harvest decay of citrus and apple

Ecogen has introduced Aspire®, a product based on a naturally occurring yeast *Candida oleophila* that is applied to disease–free fruit forcontrol of post-harvest rot pathogens of citrus, pome fruits, grapes and apple by protecting surface wounds from invasion by pathogenic fungi. It suppresses decay formation via competition with germinating spores of prominent pathogens (e.g. *Penicillium* spp.) for space and nutrients. The combination of Aspire® with 200µg/ml thiabendazole often reduced the incidence of decay, caused by the green and blue molds (*Penicillium digitatum* and *P. italicum*, respectively), as effectively as conventional fungicide treatment. Furthermore, Aspire was found high efficacious against sour rot caused by *Goetrichum condidum*, a decay not controlled by the conventional treatments (Droby et al., 1998).

Cut flowers are often subjected to pathogen infections, during shipment when high humidity has to be maintained. Even slight temperature changes can lead to condensation of water on the surface of petals. Therefore, post-harvest treatment of cut flowers is an attractive option for application of antagonists. Redmond et al. (1987) found the yeast *Exophiala jeanselmi* very effective in preventing infections by *B. cinerea* on cut roses, and found that it gave comparable results to ipridione in disease suppression (63%).

6   Biological Control with *Trichoderma* Species

One approach to biocontrol has been the use of antagonistic microorganisms that compete with, or directly attack the pathogen. Saprophytic fungi in the genus *Trichoderma* have been used to control a wide range of plant pathogenic fungi responsible for the most important diseases suffered by crops of major economic importance worldwide. In particular, isolates of *Trichoderma harzianum*, *T. virens* and *T. hamatum* have been used with success against soil-borne, seed-borne diseases and diseases in the phyllosphere and against storage rots (Tronsmo and Hjeljord, 1998). The mycoparasitic activity of *Trichoderma* spp. is due to three antagonistic mechanisms that may be operative for the control effects, including competition, antibiosis, production of cell wall-degrading enzymes or a combination of these activities.

Mycoparasitism, the phenomenon of one fungus parasitizing another, relies on the production of lytic enzymes by the mycoparasite for the degradation of cell walls of the host fungus. *Trichoderma*, a mycoparasite, can detect its host from a distance and, on detection, begins to branch in an atypical way towards the pathogenic fungus. It is speculated that trophic response is induced by nutrient gradients arising from the host (In addition, some strains of *Trichoderma* may produce non-volatile antibiotics that inhibit and presumably, predispose host hyphae to infection before contact occurs). Following recognition of its host, *Trichoderma* attaches to it and then either grows along

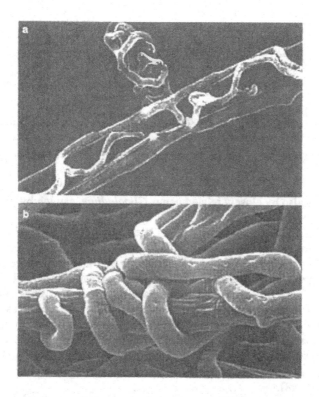

Figure 7.7 Scanning electron micrographs of interactions between *Trichoderma harzianum* and *R. Solani*. (a) *T. harzianum* penetrates the hyphae of its host and then exits (b) *T. harzianum* (thin hypha) coils around *R. solani* (thick hypha), there is a loss of turgor and marked cell lapse in *R. solani* (Benhamou and Chet, 1993).

the host hyphae or coils around them (Figure 7.7). Secretion of lytic enzymes, chitinases and other hydrolases follow. *Trichoderma* produce appressorium like bodies, which aid penetration of the host cell wall (Goldman et al., 1994). Subsequent degenerative events in the host include disorganization of cell wall structure resulting in osmotic imbalances followed by intracellular disruption, such as retraction of plasma membrane and aggregation of the cytoplasm (Figure 7.8).

The majority of pathogenic fungi contain chitin and 1, 3-β-glucans in their cell walls, and dissolution of, or damage to, these structural polymers has adverse effects on the growth and differentiation of such fungi. Cell-wall-degrading enzymes, particularly, chitinolytic enzymes produced by *Trichoderma*

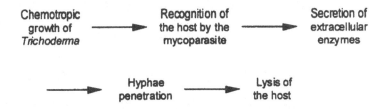

Figure 7.8   Mechanism of mycoparasitic activity of *Trichoderma* spp.

species, as also by closely related genus *Gliocladium,* can inhibit proliferation of plant pathogenic fungi.

Chitinolytic enzymes have been implicated in biological control mechanisms. In *in vitro* experiments, it has been shown that chitinolytic and gluconolytic enzymes isolated from *T. harzianum* inhibit spore germination and hyphal (germ tube) elongation in several plant pathogens, including *Botrytis cinerea, Fusarium solani, F. graminearum, Ustilago avenae* and *Uncinula necator.* However, *Pythium* spp., that does not have chitin as a structural component of cell wall was not affected. The degree of inhibition was found proportional to the level of chitin in the cell wall of target fungi. Combinations of the purified enzymes such as glucan 1,3-β–glucosidases and chitinolytic enzymes resulted in a synergistic increase in antifungal activity (Tronsmo and Hjeljord, 1998).

Zimand et al. (1996) reported a new mechanism of biological control of *T. harzianum* T 39 on *Botrytis cinerea.* Besides the inhibitory effect on germination and germ tube elongation of conidia of *B. cinerea, T. harzianum* T 39 reduced the production and activities of pathogenic pectolytic enzymes three days after inoculation, resulting increased accumulation of pectic enzyme product (i.e. oligogalachuronides). These sugars can elicit the host plant defense mechanisms, thus slowing the disease severity.

Trichoderma harzianum has been successfully applied in the phyllosphere for the control or low disease incidence of *Botrytis cinerea,* one of the most serious pathogens of grapevines. However, the time of treatment is important for control. Most effective protection is obtained by treatments extending from the time of flowering to three weeks before harvest. The best control is reported to be obtained in an integrated program together with a reduced dosage of fungicides. Even reduced amounts of the fungicide can stress and weaken the pathogen and render its propagules more susceptible to subsequent attack by the antagonist.

Similarly, *Trichoderma* spp. have been used as seed coatings and as granules for biological control of soil-borne pathogens. Control of *Pythium* spp.

in tobacco, sugar beet and cauliflower has been achieved by *T. harzianum* through soil applications (Mukhopadhyay et al., 1986). As a seed protectant, its efficacy may depend on the relative rates with which the antagonist and the pathogen colonize the seed. However, here also integrated biological-chemical seed treatments may provide better results than treatment with either the chemical or biological agent alone. While the chemical protectant can be ideal for short term protection of the seed or seedling, a rhizosphere-competant bioprotectant can colonize the entire root system and provide a season long protection unattainable through acceptable levels of chemical treatment.

Trichoderma biocontrol agents have also successfully controlled post-harvest diseases. Carrots naturally infected with *Mycocentrospora acerina* or *Rhizoctonia carotae* were treated with conidia of the cold tolerant isolate *T. harzianum* P1 at the time of harvest and stored in the plastic-wrapped bins at 0-5°C. The cold tolerant *Trichoderma* isolate significantly reduced the number of roots infected by both pathogens. On average, the amount of marketable crop increased by 47% after 6 months in cold storage, as compared to untreated controls (Tronsmo, 1989).

7 Biological Control with Fluorescent Pseudomonads

Fluorescent pseudomonads are the most frequently used plant growth promoting rhizobacteria that function by suppressing detrimental rhizosphere microflora present in most soils. *Pseudomonas fluorescens* and *P. putida* suppress several major plant pathogens, especially the soil-borne ones (Laha et al., 1996). Production of anti-fungal metabolites, such as antibiotics and hydrogen cynamide and siderophore-mediated iron competition are primary mechanisms by which these bacteria suppress disease (Weller, 1988). Siderophores are biosynthetic compounds that are produced under iron-limiting conditions. They serve to chelate the ferric ion (Fe^{3+}) from the environment into microbial cells. Thus, iron availability is reduced for the pathogens.

Pseudomonas fluorescens 2-79 has been effective in control of 'take-all' of wheat caused by *Gaemannomyces graminis* var. *tritici*, one of the most economically-important crown and root rot diseases of wheat worldwide. Aggressive root colonization is important for effective biocontrol with bacteria and is positively correlated with take-all suppression. The take-all decline, a spontaneous diminution in disease and increase in yield that occurs after several years of wheat monoculture is well known natural phenomenon associated with the preponderance of antagonistic organisms especially the fluorescent *Pseudomonas*. The suppression of take-all of wheat depends on the ability of bacteria to colonize the roots and production of an antibiotic phenazine-1-carboxylic acid (PCA), a siderophore called pyoverdine and an anti-fungal factor, 2,4-diacetyl phloroglucinol (Thomashow and Weller, 1990). However, antibiotic rather than siderophore production has been suggested to the main mode of action of *P. fluorescens* 2-79 in its control of *G. graminis* var. *tritus* (Hamadan et al., 1991).

Fluorescent species of *Pseudomonas* species have been used to control disease of various crops. Suppression of damping-off of cotton caused by *Rhizoctonia solani* and *Pythium ultimum* has been reported by applying *P. fluorescens* isolated from the rhizosphere of cotton seedling. It has been suggested that sidorophores may not be the only key factors associated with disease suppression. The antibiotics pyrrolnitrin and pyoluteorin produced by *P. fluorescens* were the key antagonistic factors against *R. solani* and *Pythium* spp. respectively. Suppression of several vescular wilts caused by *Fusarium oxysporum*, control of root rot of wheat caused by *Pythium ultimum* and potato seed piece decay by *Erwinia carotaova* by *Pseudomonas* spp., are also reported.

Novartis AG has reported development of genetically modified *Pseudomonas* strains with enhanced biocontrol activity against plant pathogenic fungi such as species of *Rhizoctonia* and *Pythium* (Ligon et al., 1999). The modified strains produce enhanced amounts of antifungal metabolites such as pyrrolnitrin that are active against these fungal pathogens.

8   Production of Biofungicides

Liquid cultivation in batch-stirred tank reactors is the standard method of producing microbial products. However, biofungicides production represents new territory in the progression of industrial fermentation technology. The biocontrol agents must not only be produced in high yield but also should have high retention of cell viability with maintenance of crop compatibility and bioefficacy during several months of storage. Medium optimization must consider not only for spore yield but also for spore fitness based on qualities such as desiccation tolerance, stability as a dry preparation and biocontrol efficacy. Nutritional and environmental conditions during culture growth and sporulation, which promote the accumulations of appropriate endogenous reserves, may be a critical factor in determining spore fitness. Once harvested, from culture, they must be stored, preferably in a dry, non-refrigerated state until time for field application. Even after storage, the liquid culture-produced cells must remain not only viable but active enough to rapidly colonize and establish plant disease protection. The biocontrol application must be free of metabolites that are phytotoxic and may have a detrimental effect on the crop (Slininger et al., 1998).

Liquid fermentation is the preferred manufacturing process for biocontrol agents. While fermentation of bacteria and yeast is straight forward and easily accomplished, this fermentation strategy presents difficult challenges for fungi. In their natural environments, fungi grow along the surfaces or within the structures of solid substrates and exhibit many different types of growth phases and morphologies. When environmental conditions are favorable and sufficient amounts of nutrients are available, fungal growth occurs primarily by the development of mycelia. However, when nutrients become limiting or environmental conditions are harsh, the fungal mycelia stop growing and form spores. Fungal growth in liquid media, particularly in fermentation vessels, is vastly different from grown on solid substrate. The mycelia have different morphological properties and have a tendency to form clumps or pellets. The occurrence of pellets greatly reduces the efficiency of fermentation and the final spore yields. Although, the vegetative fungal mycelia can be used as an active ingredient in product formulations, the mycelia can not easily be formulated or stored for extended periods of time. Consequently, many efforts have been placed on developing fermentation processes which maximize fungal spore production (Stowell, 1991). Harman and coworkers of TGT Inc. (Geneva, NY) successfully developed a large-scale fermentation technology combining liquid fermentation with a secondary solid fermentation. It resulted in an inexpensive, highly active biomass with a good shelf life. This method is now used in the commercial production of biofungicide *T. harzianum* KRL-AG22 (BioTrek 22G®) (Tronsmo and Hjeljord, 1998).

9   Formulation of Biofungicides

Intact-cell based biofungicides require very delicate and rather sophisticated approaches in the design of commercial formulation. It is necessary that the formulation ensure product stability during storage (i.e. shelf life), as well as reliability during application. The formulation of biofungicides has been a subject matter of two recent review articles (Lumsden et al., 1995; Fravel et al., 1998).

Spores are natural survival vessels that can be readily dried and re-suspended in liquid delivery systems at the time of use, such as those of *Bacillus* species. Some non-spore forming microorganisms such as pseudomonads, do not possess the survival mechanisms of spores. Thus, drying and rehydration is not possible with non-spore formers and offers formulation challenges especially with regard to stability. Therefore, lack of successful, viable formulations of fluorescent pseudomonads remain a major obstacle for their large-scale use.

9.1   *Formulation of Fungal Antagonists*

Commercial preparations of the fungus *Gliocladium virens* (GlioGard®) appeared in the market as an alginate prill formulation. Subsequently, this formulation was discontinued due to certain quality control problems. The modified formulation (SoilGard®) produced by fluid-bed granulation included dextrin as a binder and reduced content of alginate. It was reported to be stable and highly effective product for controlling damping-off of various ornamental and vegetable transplants caused by the pathogens *Pythium ultimum* and *Rhizoctonia solani* (Lumsden, et al., 1996). Alginate gel has also been used successfully to prepare formulations of biocontrol bacteria as well as other fungi, including pseudomonads, *Trichoderma* spp., *G. virens* etc. (Marois et al., 1989).

Ecogen has adopted a proprietary formulation technology to incorporate the yeast *Candida oleophila* into a water dispersible granule formulation (Aspire®) that provides stability during storage as well as user-friendliness to a product. On the other hand, *Ampelomyces quisqualis*, a pycnidial parasite of powdery mildew fungi is formulated as a wettable powder (AQ10®). Non-pathogenic *Fusarium oxysporum* is produced as a microgranular product (Fusaclean®). Similarly, *Pythium oligandrum* has been formulated as a granular or powder product (Polygandron®) for use as a seed treatment on sugarbeet for control of pathogenic *Pythium* spp.

9.2   *Formulation of Bacterial Antagonists*

Bacterial antagonists have been formulated in a variety of ways to control plant pathogens. The sporulating gram-positive bacteria offer biological solutions to the formulation problems that have plagued biocontrol. Kodiak®, a wettable powder formulation of *Bacillus subtilis* strain GBO 3, is highly effective for control of the pathogens *Fusarium* and *Rhizoctonia*, as well as in stimulating plant growth. The gram-positive microorganism offers heat- and desiccation-resistant spores that can be formulated readily into stable dry powder products. On the other hand, non-spore-forming organisms are more difficult to formulate, as they do not have the survival mechanisms of spores. The gram-negative microorganisms such as *Pseudomonas syringae* or *Burkholderia cepacia* have a short shelf life and are readily killed by desiccation. These microorganisms are traditionally formulated into various solid carriers. The different strains of *P. syringae* are formulated as wettable powder formulations (BioSave®10, BioSave®11) for post-harvest application to citrus and pome fruit. *Pseudomonas fluorescens* strain A 506 is supplied as a powder (BlightBan® A 506) for spraying onto leaf surfaces of apple and pear trees to provide protection from frost and from *Erwina amylovora*, cause of fireblight. Mycostop®, based on *Streptomyces griseoviridis* strain K 61 contains mycelium and spores and is also formulated as a wettable powder. It controls damping off and root and basal rot diseases of ornamentals and vegetables caused by *Fusarium, Pythium* and *Phomopsis.*

In view of the user-friendly nature of liquid formulations, newer developments are being made in liquid media - either aqueous or mineral oil. These formulations allow for slow, continual growth of the organism in the liquid or suspend growth to a starved level (Wacek, 1997). Examples include, *Pseudomonas fluorescens*, formulated as an aqueous suspension of fermenter biomass (Conquer®, Victus®), sprayed into mushrooms to prevent blotch caused by *Pseudomonas tolassil*. Similarly, *B. cepacia* is formulated as a liquid suspension (Deny®) for seedling drench or use as drip irrigation for a few high-value crops, such as strawberry and melons.

10   Concluding Remarks

The promise of biologicals in large-scale agricultural market is beginning to be realized, at least partially. It would be unrealistic to expect that biological control agents can completely replace chemical fungicides in disease control. There are, however, areas in which biological control is superior to chemical fungicide control. Presently, biofungicides occupy a niche market and provide reasonable level of control as part of an integrated disease management strategy. For biological control methods to emerge as an economically viable option, their consistency and efficacy in controlling soilborne and foliar fungal diseases and post-harvest decay, needs to be enhanced to a level comparable to that of synthetic fungicides. More knowledge is needed to understand the complex modes of action of the antagonistic strains and to apply them in the best conditions to achieve optimal biological control of fungal diseases in the plants. One way to improve efficacy and consistency of biological control would be to mimic the complexity of the mechanisms operating in suppressive soils, and to use several populations of antagonistic microorganism together. Improvement of field performance of biofungicides to display curative or eradictive activity is also necessary, to find commercial usage. For this, a strong collaboration and understanding between the agriculture, industry and industrial microbiologists are required to continue the advances of new biologicals.

References

1.   Alabouvette, C., B. Schippers, P. Lemanceau and P. A. H. M. Bakker, 1998. "Biological control of *Fusarium* wilts: Towards development of commercial

products", in *Plant-Microbe Interactions and Biological Control*, G. J. Boland and L. D. Kuykendall, eds., Marcel Dekker, Inc., New York, NY. pp. 15-36.

2. Backmam, P. A., M. Wilson and J. F. Murphy, 1997. "Bacteria for biological control of plant diseases", in *Environmentally Safe Approaches to Crop Disease Control*, N. A. Rechcigl and J. E. Rechcigl, eds., CRC Press, pp. 95-109.

3. Baker, R. and G. N. Griffin, 1995. "Molecular strategies for biological control of fungal plant pathogens", in *Novel Approaches to Integrated Pest Management*, R. Reuveni, ed., CRC Press, Inc., pp. 153 – 182.

4. Beltsville Agricultural Research Center, USDA, 1998. "New biocontrol formulations that are easier to make and use", www.barc.usda.gov/psi/bpdl/recent.htm.

5. Be'langer, R. R., A. J. Dik and J. G. Menzies, 1998. "Powdery mildews: recent advances towards integrated control", in *Plant-Microbe Interactions and Biological Control*, G. J. Boland and L. D. Kuykendall, eds., Marcel Dekker, New York, pp. 89-109.

6. Benhamou, N. and I. Chet, 1993. "Ultrastructure and Gold cytochemistry of the mycoparasite process", *Phytopathol.*, 83(10), 1062-1071.

7. Brannen, P. M. and D. S. Kenney, 1997. "Kodiak – a successful biological-control product for suppression of soil-borne plant pathogens of cotton", *J. Ind. Microbiol. Biotechnol.*, 19, 169 – 171.

8. Chalutz, E. and S. Droby, 1998. "Biological control of postharvest diseases", in *Plant-Microbe Interactions and Biological Control*, G. J. Boland and L. D. Kuykendall, eds., Marcell Dekker, Inc., New York, NY. pp 157-170.

9. Cook, R. J., 1993. "Making greater use of introduced microorganisms for biological control of plant pathogens", *Annl Revs of Phytopathol.*, 31, 53-80..

10. Droby, S. L., A. Cohen, B. Daus, B. Weiss, B. Horev, E. Chaletz, H. Katz, M. Koren-Tzur and A. Shanchnai, 1998. "Commercial testing of Aspire: a yeast preparation for the biological control of post-harvest decay of citrus", *Biol. Control*, 12(2), 97-101.

11. El-Ghauth, A., 1997. "Biologically-based alternatives to synthetic fungicides for the control of postharvest diseases", *J. Ind. Microbiol. Biotechnol.*, 19, 160 – 162.

12. Emmert, E. A. B. and J. Handelsman, 1999. "Biocontrol of plant disease: a (gram-) positive perspective", *FEMS Microbiol. Letts.*, 171, 1-9.

13. Erwin, D. C. and O. K. Riberio, 1996. "Cultural and biological control", in *Phytophthora Diseases Worldwide*, The American Phytopathological Society, St Paul, MN, APS Press, pp. 145-184.

14. Estrella, A. H. and I. Chet, 1998. "Biocontrol of bacteria and phytopathogenic fungi", in *Agricultural Biotechnology*, A. Altman, ed., Marcel Dekker, Inc., New York, NY, pp. 263-282.

15. Fravel, D. R., 1998. "Use of *Sporidesmium sclerotivorum* for biological control of Sclerotial plant pathogens", in *Plant-Microbe Interactions and Biological Control*, G. J. Boland and L. D. Kuykendall, eds., Marcel Dekker, Inc., New York, NY., pp., 37-47.

16. Fravel, D. R., W. J. Connick, Jr., and J. A. Lewis, 1998. "Formulation of microorganisms to control plant diseases", in *Formulation of Microbial*

Biopesticides: Beneficial Microorganisms, Nematodes and Seed Treatments, H. D. Burges, ed., Kluwer, Dordrecht, The Netherlands, pp. 187-201.

17.   Froyd, J. D., 1996. "Can synthetic pesticides be replaced with biologically-based alternatives? An industry perspective", 1996 SIM Annual Meeting, Aug 4-6, 1996, Abstracts, S41.

18.   Goldman, G. H., C. Hayes and G. E. Harman, 1994. "Molecular and cellular biology of biocontrol by *Trichoderma* spp.", *TIBTECH*, *12*, 478-482.

19.   Hamdan, H., D. M. Weller and L. S. Thomashow, 1991. "Relative importance of fluorescent siderophores and other factors in biological control of *Gaeumannomyces graminis* var. *tritici* by *Psudomonas fluorescens* 2-79 and M4-80R, *Appl. Environ. Microbiol.*, 57, 3270-3277.

20.   Harman, G. E. and E. B. Nelson, 1994. "Mechanisms of protection of seed and seedlings by biological seed treatments: implications for practical disease control", *Seed Treatments: Progress and Prospects*, T. Martin, ed., British Crop Protection Council, Farnham, UK, pp. 283-292.

21.   Hofstein, R. and A. C. Chapple, 1999. "Commercial development of biofungicides", in *Biopesticides: Use and Delivery*, F. R. Hall and J. J. Menn, eds., Humana Press, Totowa, NJ, pp. 77-102.

22.   Jones, D. A. and A. Kerr, 1989. "*Agrobacterium radiobacter* strain K1026, a genetically engineered derivative of strain K84, for biological control of crown gall", *Phytopathol.*, 73, 15-18.

23.   Ko, W. H., 1982. "Biological control of phytophthora root rot of papaya with virgin soil", *Plant Disease*, 66(6), 446-448.

24.   Kohl, J. and N. J. Fokkema, 1998. "Strategies for biological control of Necrotrophic fungal foliar pathogens", in *Plant-Microbe Interactions and Biological Control*, G. J. Boland and L. D. Kuykendall, eds., Marcell Dekker, Inc., New York, NY. pp. 49-88.

25.   Laha, G. S., R. P. Singh, and J. P. Verma, 1996. "Role of growth promoting rhizobacteria in plant disease management", in *Disease Scenario in Crop Plants, Vol. II - Cereals, Pulses, Oilseeds and Cash Crops*, V. P. Agnihotri, O. Prakash, R. Kishun and A., K. Misra, eds., Int'l Books and Periodicals, Delhi, India. pp. 233-241.

26.   Laing, S. A. K. and J. W. Deacon, 1991. "Video microscopical comparison of mycoparasitism by *Pythium oligandrum*, *P. nunn* and an unnamed *Pythium* species", *Mycol. Res.*, 95, 469-479.

27.   Lemanceau, P. and C. Alobouvette, 1993. "Suppression of fusarium-wilts by fluorescent pseudomonads: mechanisms and applications, *Biocontrol Sci. Technol.*, 3, 219-224.

28.   Lemanceau, P., P. A. H. M. Baker, W. J. DeKogel, C. Alabouvette and B. Schippers, 1992. "Effect of pseudobactin 358 production of *Pseudomonas putida* WCS 358 on suppression of fusarium wilt of carnations by nonpathogenic *Fusarium oxysporum* Fo47", Appl. Environ. *Microbiol.*, 58, 2978-2982.

29.   Lifshitz, R., M. Dapler, Y. Elad and R. Baker, 1984. "Hyphal interaction between a mycoparasite, *Pythium nunn* and several soil fungi", *Can. J. microbiol.*, 30, 1482.

30.   Ligon, J. M., N. R. Torkewitz, D. S. Hill, T. Gaffney and J. M. Stafford (Novartis AG), 1999. "Genetically modified *Pseudomonas* strains with enhanced biocontrol activity", U.S. Patent 5,955,348 (September 21, 1999).

31.   Lumsden, R. D., J. A. Lewis and D. R. Fravel, 1995. "Formulation and delivery of biocontrol agents for use against soilborne plant pathogens", in *Biorational Pest Control Agents - Formulation and Delivery*, F.R. Hall and J.W. Barry, eds., American Chemical Society, Washington, DC, pp. 166-182.

32.   Lumsden, R. D., J. F. Walter and C. P. Baker, 1996. "Development of *Gliocladium virens* for damping-off disease control", *Can. J. Plant Pathol.*, 18, 463-468.

33.   Maloney, A., 1995. "Sources for non-chemical management of plant diseases", *Adv. Plant Pathol.*, Vol. 2, pp. 104-130.

34.   Marrone, P., 1999. Personal communication.

35.   Martin, F. N. and J. G. Hancock, 1987. "The use of *Pythium oligandrum* for biological control of preemergence damping-off caused by *P. ultimum*", *Phytopathology*, 77, 1013-1020.

36.   Marois, J. J., .D. R. Fravel, W. J. Connick, Jr., H. L. Walker and P. C. Quimby, 1989. U. S. Patent 4818530.

37.   Mehrotra, R. S., K. R. Aneja and A. Aggarwal, 1997. "Fungal control agents", in *Environmentally Safe Approaches to Crop Disease Control*, N. D. Rechcigl and J. E. Rechcigl, eds., CRC Press, Boca Raton, Florida, pp. 111-137.

38.   Mukhopadhyay, A. N., 1994. "Biocontrol of soil-borne fungal plant pathogens - current status, further prospects and potential limitations", *Indian Phytopathol.*, 47(2), 119-126.

39.   Mukhopadhyay, A. N., A. Brahmbhatt and G. J. Patel, 1986. "*Trichoderma harzianum* - a potential biocontrol agent for tobacco damping off", *Tobacco Res.*, 12, 26-35.

40.   Mycotech Corporation, 2000, "New Biological Fungicide in Development", http// www.mycotech.com.new/.

41.   Orlikowski, L. B. and A. Schmidle, 1985. "On the biological control of *Phytophthora cactorum* with *Trichoderma viride*", Nachrichtenbl. Dtsch. Pflanzenschutzdienst, 37, 78-79.

42.   Redmond, J. C., J. J. Marois and J. D. MacDonald, 1987. "Biological control of *Botrytis cinerea* on roses with epiphytic microorganisms", *Plant Dis.*, 71, 799.

43.   Seddon, B., S. G. Edwards and L. Rutland, 1995. "Development of *Bacillus* species as antifungal agents in crop protection", in *Modern Fungicides and Antifungal Compounds*, 11th Int'l Symposium, Thuringia, Germany, H. Lyr, P. E. Russel and H. D. Sisler, Eds., Intercept Ltd., Andover, Hampshire, U.K. pp. 555-560.

44.   Sen, B., K. Angappan and P. Dureja, 1996. "Multiprong actions of biocontrol agent, *Aspergillus niger* (AN 27)", Abstracts, 2nd Int'l Crop Sci. Congr. On *Crop Productivity and Sustainability – Shaping the Future*, Nov. 17-24, 1996, New Delhi, p. 301.

45.   Sivan, A. and G. E. Harman, 1991. "Improved rhizosphere competence in a protoplast fusion progeny of *Trichoderma harzianum*", *J. Gen. Microbiol.* 137:23-29.

46.   Slininger, P. J., J. E. Van Cauwenberge, M. A. Shea-Wilbur and R. J. Bothast, 1998. "Impact of liquid culture, physiology, environment and metabolites on

biocontrol agent qualities", in *Plant Microb Interactions and Biological Control*, G. S. Boland and l. D. Kuykendall, eds., Marcel Dekker, New York, pp. 329-353

47. Stowell, L. J., 1991. "Submerged fermentation of biological herbicides", in *Microbial Control of Weeds*, D. O. TeBeest, ed., Chapman & Hall, N. Y., pp. 225-261.

48. Thomashow, L. S. and D. M. Weller, 1990. "Role of antibiotics and siderophores in biological control of take-all disease of wheat", *Plant Soil*, 129, 93-99.

49. Tronsmo, A. and L. G. Hjeljord, 1998. "Biological control with *Trichoderma* species", in *Plant-Microbe Interactions and Biological Control*, G. S. Boland and L. D. Kuykendall, eds., Marcel Dekker, New York, pp. 111-124.

50. Tronsmo, A. 1989. "*Trichoderma harzianum* used for biological control of storage rot on carrots", *Norwegian J. Agric. Sci.*, 3, 157-161.

51. Utkhede, R. S., 1996. "Potential and problems of developing bacterial biocontrol agents", *Can. J. Plant Pathol.*, 18, 455-462.

52. Utkhede, R. S., 1986. "Biology and control of apple crown rot caused by *Phytophthora cactorum*", *Phytoprotection*, 67, 1-13.

53. Van Driesche, R. G. and T. S. Bellows, Jr., 1996. *Biological Control*, Chapman & Hall, pp. 255 – 256.

54. Verhaar, M. A., P. A. C. van Strien and T. Hijwegen, 1993. "Biological control of cucumber powdery mildew (*Sphaerotheca fuliginea*) by *Verticillium lecanii* and *Sporothrix cf. Flocculosa*", in Biological Control of Foliar and Postharvest Diseases, N. J. Fokkema, J. Kohl and Y. Elad, eds., IOBC/WPRS Bulletin 16(11), 79.

55. Wacek, T. J., 1997. "Liquid formulations of non-spore forming micro-organisms, in *Pesticide Formulations and Application Systems: 17th Volume*, G. R. Goss, M. J. Hopkinson and H. M. Collins, eds., American Society for Testing and Materials, pp. 94-97.

56. Weller, D. M., 1988. "Biological control of soil-borne plant pathogens in the rhizosphere with bacteria, *Annu. Rev. Phytopathol.*, 26, 379-407.

57. Whipps, J. M., 1997. "Developments in the biological control of soil-borne plant pathogens", in *Advances in Botanical Research, Volume 26*, J. A. Callow, ed., Academic Press, London, *pp.* 1-134.

58. Wilhite, S. E., R. D. Lumsden and D. C. Straney, 1994. "Mutational analysis of gliotoxin production by the biocontrol fungus *Gliocladium virens* in relation to suppression of *Pythium* damping-off", *Phytopathol.*, 84, 816-821.

59. Wilson, C. L., 1997. "Biological control and plant diseases – a new paradigm", *J. Ind. Microbiol. Biotechnol.*, 19, 158-159.

60. Wilson, M., 1997. "Biocontrol of aerial plant diseases in agriculture and horticulture: Current approaches and future prospects", *J. Ind. Microbiol. Biotechnol.*, 19, 188 – 191.

61. Zimand, G., Y. Elad and I. Chet, 1996. "Effect of *T. harzianum* on *Botrytis cinerea* pathogenicity", *Phytopathlogy*, 86, 1255-1260.

62. Zhou, T. and G. J. Boland, 1998. "Biological control strategies for *Sclerotinia* diseases", in *Plant-Microbe Interactions and Biological Control*, G. J. Boland and L. D. Kuykendall, eds., Marcell Dekker, Inc., New York, NY., pp. 127-156.

8

Bioherbicides

1   Introduction

The use of herbicides with documented adverse affects such as residues in ground water and persistence of some herbicides beyond a single growing season affecting rotational crops have led to increased interest in alternatives such as biological weed control. Biological herbicides offer yet another beneficial trait, *i.e.* a killing action that is different from synthetic herbicides, so they should be efficient tools in combating weed resistance to specific chemicals.

Hoagland (1996) defined bioherbicides as plant pathogens, phytotoxins derived from pathogens or other microorganisms applied to control weeds. The whole foundation of biological weed control is to push the disease process by tipping the ecological balance in favor of the pathogen. This is often done to identify the 'right' pathogen or the one that will work in diverse environments.

A pathogen could qualify as the active ingredient for a commercial product only if it can be applied in a viable form, at inoculum levels high enough to initiate an infection and by manipulating the micro-environment for a long enough period of the time. It is required to make sure that an infection gets to the point that it can perpetuate itself. This would involve developing formulations that allow them to work consistently under diverse field conditions, including stability of living organisms that would survive a distribution process for as long as 18 months between packaging and use.

The majority of bioherbicide candidates studied are the fungal pathogens. Nevertheless, recently it has been shown that it is possible to consider the use of bacteria also as bioherbicide. The future potential of mycoherbicides, in particular, is seen in areas that are served inadequately by chemical herbicides. These areas include (1) control of parasitic weeds; (2) control of weeds closely related to crops (crop mimics), in which case a high degree of selectivity is necessary; (3) control of weeds resistant to chemical herbicides and (4) control of weeds infecting small, specialised areas where development of chemical herbicides would be too costly (Templeton et al., 1986).

Several reviews on bioherbicides have appeared in the literature recently (Mortensen, 1998; Hoagland, 1996; Christy et al., 1993, Gardner and McCoy, 1992 and Zorner et al., 1992). In this chapter, an overview on various bioherbicides that have been registered and commercialised is provided.

2   Fungal Bioherbicides

Fungi are capable of entering the plant host through wounds, through openings in the epidermis (for example, stomata, nectaries), or by direct penetration of the cuticle by germinating spores. Development of phytopathogenic fungi as mycoherbicides has progressed well. Charudattan (1991) has cited identification and development of over 109 pathogens, out of which 73 organisms having been placed in commercial development programs in the decade of 80's. Self-disseminating pathogens (such as rusts, smuts) as well as poorly disseminating pathogens, such as soil-borne and mucoid-spored fungi, have been explored.

Bioherbicidal pathogens being studied presently are host-specific for reasons of environmental safety. Extreme host-specificity is an advantage where a weed is closely related to the crop, in which it is to be controlled. However, there may not be enough perspective host-specific pathogens for each problem weed. Besides, use of host specific agents will be precluded for economic reason alone due to presence of multiple weed species with each crop. Therefore, only broad-host range pathogens are likely to compete well with broad-spectrum herbicides, by virtue of their targeting multiple weed species in single application (Sands and Miller, 1992). The well-known broad-host range plant pathogens include, *Sclerotinia sclerotiorum, Pythium ultimum, Sclerotium rolfsii, Phymatotrichum omnivorum* and *Pseudomonas solanacearum.* Often soil-borne, these pathogens are capable of surviving in the soil after killing one plant until they attack another susceptible plant of the same or different species. Their mode of action involves root or crown invasion, vascular plugging and wilting. However, as with other plant pathogens, the mechanisms involved in host range restriction, are unknown.

Daniel et al. (1973) noted that fungi selected as candidate mycoherbicide should possess certain essential characteristics. The fungus must be amenable to *in vitro* production. The product must also remain stable in culture as well as in storage, have no dormancy factors that limit infectivity, and be able to infect weed hosts in a relatively broad spectrum of environments. Fungi possessing these characteristics are generally facultative pathogens. However, only a few of these organisms have been commercialised (Table 8.1).

A. cassiae is effective against sicklepod (*Cassia obtusifolia*) in soybean grown in diversity of soil types and environmental conditions. It causes a foliar blight disease in host plants and is being developed by Mycogen Corporation under the trade name Casst® for sicklepod control.

Collego® was developed in a cooperative effort between the University of Arkansas, USDA and private industry. It is marketed as a complex formulation of dried conidia of *C. gloeosporioides* f. sp. *Aeschynomene*.

Table 8.1 Registered fungal bioherbicides

Pathogen	Target Weed	Crop	Trade Name	Manufacturer
Alternaria cassiae	Sicklepod (*Cassia obtusifolia*)	Soybean	Casst	Mycogen Corp. San Diego, CA
Colletotrichum gloeosporioides f. sp. *aeschynomene*	Northern jointvetch (*Aeschynomene virginica*)	Rice and irrigated soybean	Collego	Pharmacia & Upjohn Kalamazoo, MI; Encore Technologies, Minnetonka, MN
Phytophthora palmivora (Butl.)	Stragler vine (*Morrenia odorata*)	Citrus groves	DeVine	Abbott Labs. Chicago, IL, USA
Colletotrichum Gloeosporioides f. sp. *malvae*	Round-leaved mallow (*Malva pusilla*)	Vegetable crops and straw-berries	BioMal	Philom Bios Saskatoon, Canada
Puccinia Canaliculata (Schw.)	Yellow nutsedge (*Cyperus esculentus L.*)	Sugarcane, maize, potato, cotton and soybean	Dr. BioSedge	Tifton Innovation Corp., Tifton, GA

C. gloeosporioides is a facultative sporophyte that causes a lethal stem and foliage blight of its host weed when inoculated with spores. The formulation is rehydrated in aquous solutions. It is applied aerially to the surface of rice paddies, when stems of northern jointvetch begin to exceed the height of the rice, prior to initiation of the flowering of the weed. Within 2 weeks, Lesions form on the stems of the weed, followed rapidly by weed death. Collego® provides long-term control of northern jointvetch, curbing the need for periodic reapplication.

Similarly, DeVine®, the commercial bioherbicide based on *Phtophthora palmivora* is formulated as a fresh preparation of the biocontrol agent by Abbott Laboratories. *P. palmivora* is a facultative parasite that produces a lethal root rot of its host plant (strangler vines) and persists saprophytically in the soils for extended period of time. The fungus was first isolated from milkweed vine (*Morrenia odorata*) in Florida and is registered exclusively for its control in Florida citrus. DeVine® is a liquid preparation of viable chlamydospores produced in submerged culture. It is applied as a post-emergent herbicide to the soil around citrus trees to infect seedling and mature vines within 2-10 weeks after application. The infection process begins at the soil line and moves down to attack the roots, eventually overpowering even the largest strangler vines without damaging the citrus trees. Control is usually 100% and could last for over 2 years (Charudattan, 1988).

BioMal®, a fungal bioherbicide based on a *Colletotrichum gloeosporioides* f. sp. *malvae* was developed by Philom Bios, a Saskatoon biotechnology company. It infects only round-leaved mallow (*Malva pusilla*) which is a problem weed in Western Canada and is very effective at killing it. On application to the field, the fungal spores settle on the leaves of the plant and germinate growing mycelium. These thread like extensions act as little hooks, securing themselves to the outside of the leaf. From there, the fungus penetrates the leaf and infects the entire plant. The weed rots and is killed due to inability of taking up water or nutrients. The fungus eventually grows more spores that are carried by the wind to find another Round-leaved Mallow plant to infect. *C. gloeosporioides* breaks through the leaf structure of round-leaved mallow but cannot invade other plants. This is important as round-leaved mallow often grows in agricultural crops that are closely related. In this situation, the use of chemical herbicides is limited because the agricultural crops may be damaged. Applying BioMal eliminates the weeds without risk to the crop or the environment.

Nutsedges (*Cyperus* spp.) comprise a group of commonly occurring weeds that are among the most difficult to control. Chemical weed control programs are seriously inadequate to control nutsedges. Frequently, the weed germinates below the treated zone and avoids herbicide injury. An alternative is offered by the use of microbes that have herbicide activity specific for the problem weeds and do not infect desirable plants. Yellow nutsedge *(Cyperus exculentus)* is a serious or principal weed of sugarcane, maize, potato, cotton and soybean, chiefly in southern Africa and North America. Dr. BioSedge, a bioherbicide based on *Puccinia Canaliculata* (Schw.) has been reported to be effective in controlling yellow nutsedge. It has been commercialised by Tifton Innovation Corporation (Tifton, GA). Kadir and Charudattan (1999) have obtained a patent on the application of a fungal pathogen *Dactylaria higginsii* (Luttrell) M. B. ellis which is claimed to control *Cyperus* spp. including, purple nutsedge *(C. rotundus)*, yellow nutsedge *(C. esculentus)*, annual sedge, globe sedge and rice flat sedge *(C. iria)*. Purple nutsedge is particularly known to be an aggressive weed that is resistant to control measures.

Scheepens et al. (1996) of Ciba-Geigy have obtained a patent of fungal pathogen *Ascochyta caulina* acting as mycoherbicide for control of weeds of familly *Chenopodiaceae* in crops, especially annual herbaceous weed *Chenopodium album.*

These fungal bioherbicides require specific conditions for infection and disease development, such as a prolonged dew period in excess of 24 hours. Synergistic interactions of chemicals, mostly herbicides and plant growth regulators, and fungal weed pathogens have been discussed in Chapter 10. Such interactions could help provide improved bioherbicidal weed control and reduce the amount of herbicide application. For example, *Alternaria cassiae* functions as a fungal bioherbicide to control sicklepod, and shows enhanced control in presence of glyphosate. Glyphosate suppresses the defence response of the weed by lowering phytoalexin production and thus acts synergistically with the pathogen.

3 Bacterial Bioherbicides

Bacterial bioherbicides research is focused primarily on use of a rhizobacteria. In contrast to fungal bioherbicides using foliar plant pathogens, which have been mostly directed towards dicotyledonous weeds, rhizobacteria have been directed towards grassy weeds (monocotyledons) in cereal crops. Rhizosphere strain of *Pseudomonas flourescens* inhibited germination and growth of downy brome *(Bromus tectorum)* without affecting the growth of wheat. Kremer et al. (1990) demonstrated rhizobacteria representing diverse gram-negative bacterial genera, as potential biocontrol agents for broad leaf weeds.

Zorner et al. (1996) of Mycogen Corporation reported development of two gram-negative bacteria, namely *Pseudomonas syringae* pv. *togetes* (PST) and *Xanthomonas compestris* pv. *poaea* (X-PO*)*. PST demonstrated an ability to control a variety of weeds in the family *Asperogens*, while X-PO has been found to be specific to annual bluegrass (*Poa annua L.*). Neither of these organisms is dependent on a dew period for host colonisation and they both demonstrate their efficacy under field conditions. *Xanthomonas campestris*, a vascular phytopathogenic bacterium is applied in the early spring to newly mown grass. The bacterium rapidly colonises the xylem, the foliage wilts and the plant die within six weeks. The organism requires a prolonged period of warm temperature and some form of mechanical wounding to allow penetration for disease progression. The strain controls blue grass in turf but does not harm desirable turf species.

Christy et. al. (1993) reported development of combinations of bacterial and chemical agents for enhanced weed control. This approach has been termed as the X-tend bioherbicide system by Crop Genetics International, a U.S. biotechnology company developing it. A range of bacteria and herbicide combinations has been tested observing synergistic interactions in both greenhouse and field trials (see Chapter 10, Section 3.3). Nevertheless, a need for more robust strains or improved formulations is felt for commercialisation.

4   Environmental Limitations for Efficacy of Bioherbicides

Both, fungal bioherbicides, as well as bacteria based bioherbicides, suffer from environmental limitations for efficacy under field conditions. A fungus requires specific conditions for infection and disease development. Some period of dew on the surface of the plant is often required, so that the fungal spore can germinate and infect the plant. The dew periods of known fungal bioherbicides range from 8 to 72 hours. On the other hand, it may be possible to escape the limitation of a dew requirement in case of bacterial bioherbicides. The infection process occurs through movement of the bacteria into the host via a direct wound introduced by a mowing operation or some form of mechanical wounding to allow penetration in the host for colonisation. However, the bacteria also require a prolonged period of warm temperature and this may prove to be a limitation in cooler climates (Figure 8.1). Since all bioherbicides are living organisms, it is likely that all potential pathogenic agents will carry some sort of environmental limitation on efficacy, a fact that needs to be dealt with effectively.

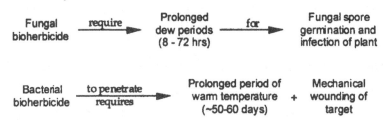

Figure 8.1   Environmental limitations of bioherbicides

5   Production of Bioherbicides

Product variability during production can be one of the reasons for unreliable field performance of a microbial herbicide. An effective biological control product would be consistently produced and have a shelf life of 1-2 years. Production and formulation are closely inter-related in the development of cost-effective inoculants. Inoculant is usually mass-produced by submerged fermentation. However, as spores are preferred for formulation, this method is used typically for those fungi that sporulate in the submerged state. Temperature, aeration and the balance between nutritional elements control sporulation of filamentous fungi in submerged culture. Controlled culture in the homogenous medium in a bioreactor is monitored and manipulated biochemically to improve efficacy of inoculant.

Most commonly, inoculants are spores, separated from medium and mycelium by filtration and centrifugation. In Collego®, 80-85% of the propagules are fission spores, 8-10% are conidia and 5% blastospores plus arthrospores (Churchill, 1982). Mycelial inoculants can be effective, as with DeVine®, but lack of shelf life precludes its widespread use. Alternatively, mycelium can be harvested and treated to induce sporulation, but this adds significantly to the cost.

The target shelf life of 1-2 years has been achieved by drying spores that have been harvested from submerged culture and mixed with inert filler such as kaolin. However, difficulties have been encountered in stabilising some fungal spores, including those of *Colletotrichum tuncatum* (Jackson et al., 1996) and *Phytophthora palmivora* (Kenney, 1986).

Some microbial herbicides have been produced on materials such as wheat, straw, oat grains and corn meal. However, propagules in solid substrate may be difficult to separate or extract for formulation. Alternatively, the substrate may become part of the final product, as with some insect pathogenic fungi. Nevertheless, solid substrate fermentation may be the only method to produce some fungi.

6   Formulation of Fungal Bioherbicides

Development of reliable and efficacious fungal bioherbicidal formulation can often be challenging. One of the goals is to keep the propagules viable and inactive for a reasonable length of time, say 1-2 years. To date, most bioherbicide formulations are concentrated on maintaining fungal agent viability in storage and reducing dew requirements. DeVine® is an aqueous concentrate of chlamydospores of *Phytophthora palmivora*. It is being used for control of stanglervine (*Morrenia odorata*) in citrus orchards in Florida. The product is not very stable and has a shelf life of only 6 weeks, when the product is refrigerated. Collego® is a two component formulation. One package component consists of dried spores of *Colletotrichum gloeosporioides f.sp. aeschynomene* in a wettable powder. A second package contains rehydrating medium. The product is reconstituted in a two-step process by first preparing the rehydrating solution then adding the dry spores and mixing well. BioMal® is a wettable powder composed of dried spores of *Colletotrichum gloeosporioides* in silica gel added directly to spray tank. Casst® is formulated as spores of *Alternaria cassiae* in emulsifiable paraffinic oil. Conidia of some fungi have proven to be difficult to stabilize in dry formulations and for these, more resistant formulations have been chosen. Among these, oil-based suspension emulsions and invert emulsions have attracted most attention.

6.1   Invert Emulsion Formulation

The biocontrol efficacy and practical usage of most fungal bioherbicides are adversely affected by lengthy dew requirements, which range from 8 to 72 hours or more. The elimination or reduction of dew requirements should improve the biological control potential of fungal bioherbicides. One approach that has shown promise in reducing dew requirements of bioherbicides, involves the use of invert emulsion formulations. An invert emulsion consists of water suspended in oil, in contrast to a standard emulsion, in which oil is suspended in water. The ability of inverts to trap water around fungal spores and retard evaporation of water spray droplets during and after application could prove beneficial for their use with bioherbicides.

Boyette et al. (1993) have studied the efficacy of an invert formulation for the pathogen *Colletotrichum truncatum* for weed control of Hemp sesbania *(Sesbania exaltata),* a perennial weed of cotton and soybean under field conditions. The oil phase of the invert emulsion consisted of parafinic oil, monoglyceride emulsifier, parafinic wax and lanolin. The aqueous phase contained conidia suspended in distilled water. The two phases were mixed at a ratio of 2:3 by volume, aqueous phase:oil phase. They reported an optimal weed control that was as good as the level of control achieved with acifluorfen, a very effective chemical herbicide (Figure 8.2) (Table 8.2).

Amsellem et al. (1990) reported similar enhanced efficacy through invert emulsion formulation of fungal bioherbicide *Alternaria cassiae* for controlling sicklepod. However, it was found that the non-pathogenic fungi were able to infect a variety of plants, when applied in the invert emulsion. They concluded that specificity of *A. cassiae* was abolished either due to physical damage to the host plant cuticle, or suppression of the host defence responses, and that an invert emulsion of *A. cassiae* in the field could cause non target damage.

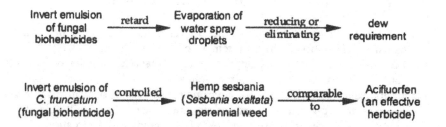

Figure 8.2 Invert emulsion formulation of fungal bioherbicides display improved biological control potential

Connick et al. (1991) reported development of an invert emulsion that exhibited lower viscosities and greater water retention properties. Also, vegetable oils could be used to enhance efficacy of fungal bioherbicides, such as *Colletotrichum orbiculare*, for control of spiny cocklebur (Auld, 1993). This invert emulsion demonstrated improved spreading properties and no phytotoxicity.

6.2 Oil-based Suspension Emulsions

Oil suspension emulsions of bioherbicides have been investigated as less expensive, easy to prepare alternatives to oil invert emulsion formulations, that could be applied with conventional spray equipment and effectively used at relatively reduced volumes. Similar formulations have been very effective for *Btk* at low volume application and gave lower rate of evaporation. Egley and Boyette (1995), however found that suspension emulsions were inferior to invert emulsions as a water source for bioherbicide activity, in the absence of a dew period.

Table 8.2 Control of Hemp sesbania with *C. truncatum* and Soybean yield under field conditions

Spray treatment	Carrier rate (l/ha)	Hemp sesbania (% mortality)	Soybean yield (kg/ha)
Conidia / water	94	19	1400
	187	40	2160
Conidia / Invert	94	72	2181
	187	97	2593
Acifluorfen	187	98	2618

7   Formulation of Bacterial Bioherbicides

One of the challenges confronting the use of plant pathogenic bacteria for
biological weed control is the requirement of free water for dispersal and the
need for wounds or natural openings, such as stomata, hydathodes or lenticels,
for entry of the bacteria into the plant (Johnson et al., 1996). The majority of
plant pathogenic bacteria are more environmentally sensitive than most fungi,
which often have pigmented spores or structures that are naturally adapted to
withstanding environmental stress, such as UV radiation and desiccation.

Zidack et al. (1992) reported that organosilicon surfactant Silwet L-77
facilitated the penetration and entry of bacteria directly into the weed (without
wounding) via open plant stomata and hydathodes. A low surface tension, 30
dynes/cm or lower is required for penetration of liquid into the stomata of a leaf.
Silwet reduces the water surface tension to 20 dynes/cm. Formulation of
Pseudomonas syringae pv. Tagetis (PST) with this surfactant resulted in
significant increases in disease severity and incidence in perennial weed such as
Canada thistle (*Cirsium arvense*), when compared to the bacterial formulation
without the surfactant. It has been suggested that delivery of bacteria into these
natural openings protect them from UV irradiation and desiccation.

8   Concluding Remarks

The commercialization of mycoherbicides such as DeVine® and Collego®
illustrates the potential of the use of phytopathogenic agents as bioherbicides to
control selected weeds. However, bioherbicides are viewed as complimentary
adjuvants to current weed management practices rather than as alternatives to
chemical herbicides. It has been suggested that development of contained broad-
host range pathogens might better compete with broad spectrum herbicides by
virtue of their use against multiple target weeds, reduced residue risk and
possibly 'perceived' aspects of safety (Sand and Miller, 1993).

Further research on formulation of fungal bioherbicides to overcome or
reduce dew requirements is also necessary. A need for shorter dew periods can
improve reliability and efficacy, as well as reduce inoculant dose requirement
(Amsellem et al., 1990).

References

1.   Amsellem, Z., A. Sharon, J. Gressel and P. C. Quimby, Jr., 1990. "Complete
        abolition of high inoculum threshold of two mycoherbicides (*Alternaria
        cassiae* and *A. crassa*) when applied in invert emulsion", *Phytopathology, 80*,
        925-929.

2.   Auld, B. A., 1993. "Vegetable oil suspension emulsions reduce dew dependence of a mycoherbicide", *Crop Protect.*, 12, 477-479

3.   Boyette, C. D., P. C. Quimby, Jr., C. T. Bryson, G. H. Egley and F. E. Fulgham, 1993. "Biological control of hemp sesbania *(Sesbania exaltata)* under field conditions with *Colletotrichum truncatum* formulated in an invert emulsion", *Weed Sci.*, 41, 497-500.

4.   Boyette, C. D., P. C. Quimby, Jr., A. J. Ceasar, J. L. Birdsall, W. J. Connick, Jr., D. J. Daigle, M. A. Jackson, G. H. Egley and H. K. Abbas, 1996. "Adjuvants, formulation and spraying systems for improvement of mycoherbicide", *Weed Technol.*, 10(3), 637-644.

5.   Charudattan, R., 1988. "Management of pathogens and insects for weed control in agroecosystems, in *Fungi in Biological Control Systems*, ed., Manchester University Press, Menchester, pp. 86-110.

6.   Charudattan, R., 1991. "The mycoherbicidal approach with plant pathogens", in *Microbial Control of Weeds*, D. O. TeBeest, ed., Chapman Hall, New York, pp. 24-37.

7.   Christy, A. L., K. A. Herbst, S. J. Kostka, J. P. Mullen and P. S. Carlson, 1993. "Synergizing weed biocontrol agents with chemical herbicides", in *Pest Control with Enhanced Environmental Safety*, S. O. Duke, J. J. Menn and J. R. Plimmer, eds., American Chemical society, Washington, D.C., pp. 87-100.

8.   Churchill, B. W., 1982. "Mass production of microorganisms for biological control, in *Biological Control of Weeds and Plant Pathogens*, R. Charudattan and H. L. Walker, eds., Wiley, New York, pp. 139-156.

9.   Connick, W. J., Jr., D. J. Daigle, C. D. Boyette, K. S. Williams, B. T. Vinyard and P. C. Quimby, Jr., 1996. "Water activity and other factors that affect the viability of *Colletotrichum truncatum* in wheat flour-kaolin granules ('Pesta')", *Biocontrol Sci. Technol.*, 6, 277-284.

10.   Connick, W. J., Jr., D. J. Daigle and P. C. Quimby, Jr., 1991. "An improved invert emulsion with high water retention for mycoherbicide delivery", *Weed Technol.*, 5, 442-444.

11.   Daniel, J. T., G. E. Templeton, R. J. Smith and W. T. Fox, 1973. *Weed Sci.*, 21, 303-307.

12.   Egley, G. H. and C. D. Boyette, 1995. "Water-corn oil emulsion enhances conidia germination and mycoherbicidal activity of *Colletotrichum truncatum*", *Weed Sci.*, 43, 312-317.

13.   Green, S., S. M. Stewart-Wade, G.J. Boland, M.P. Teshler and S.H. Liu, 1997. "Formulating microorganisms for biolological control of weeds", in *Plant-*

14.   *Microbe Interactions and Biological Control*, G. J. Boland and L. D. Kuy Kendall, eds., Marcel Dekker, Inc., New York, N.Y., USA, pp. 249-281.

15.   Gardner, W. A. and C. W. McCoy, 1992. "Insecticides and herbicides", in *Biotechnology of Filamentous Fungi; Technology and Products*, D. B. Finkelstein and C. Ball, eds., Butterworth – Heinemann, Stoneham, MA, USA, pp. 335-359.

16.   Hoagland, R. E. 1996. "Chemical Interactions with bioherbicides to improve efficacy", *Weed Technol.*, *10*, 651-674.

17.   Jackson, M. A., D. A. Schisler, P. J. Slininger et al., 1996. "Fermentation strategies for improving the fitness of a bioherbicide", *Weed Tech.*, 10, 645-650.

18.   Johnson, D. R., D. L. Wyse and K. J. Jones, 1996. "Controlling weeds with phytopathogenic bacteria", *Weed Technol.*, 10, 621-624

19.   Kadir, J. and R. Charudattan, 1999. "Control of *Cyperus* spp. with a fungal pathogen", U.S. Patent 5,945,378 (August 31, 1999).

20.   Kremer, R. J., M. F. T. Begonia, L. Stanley, E. T. Lanham, 1990. "Characteristics of rhizobacteria associated with weed seedlings", *Appl. Environ. Microbiol.* 56, 1649-1655.

21.   Pilgeram, A. L. and D. C. sands, 1998. "Mycoherbicides", in *Biopesticides: Use and delivery*, F. R. Hall and J. J. Menn, eds., Humana Press, Totowa, N. J., pp. 359-370.

22.   Mortensen, K., 1998. "Biological control of weeds using microorganisms, in *Plant Microbe Interations and Biological Control*", G. J. Boland and L. D. Kuy Kendall, eds., Marcel Dekker, New York, pp. 223-248.

23.   Sands, D. C., and R. V. Miller, 1993. "Altering the host range of mycoherbicides by genetic manipulation", in *Pest Control with Enhanced Environmental Safety*, S. O. Duke, J. J. Menn and J. R. Plimmer, eds., American Chemical Society, Washington, D.C. pp. 101-109.

24.   Templeton, G. E., R. J. Smith Jr., and D. O. TeBeest, 1986. "Progress and potential of weed control with mycoherbicides", *Rev. Weed Sci.*, 2, 1-14.

25.   Zidack, N. K., P. A. Backman and J. J. Shaw, 1992. "Promotion of bacterial infection of leaves by an organosilicon surfactant: Implications for biological weed control", *Biol. Control*, 2, 111-117.

25.   Zorner, P. S., S. L. Evans and S. D. Savage, 1992. "Perspectives on providing a realistic technical foundation for the commercialisation of bio-herbicides", in *Pest Control with Enhanced Environmental Safety*, S. O. Duke, J. J. Menn and J. R. Plimmer, eds., American Chemical Society, Washington, D.C., pp.79-86.

26.   Zorner, P. S., S. D. Savage, S. L. Evans and P. Simpson, 1996. "Bacteria as biological herbicide", Abstracts, *Soc. Industr. Microbiol. Meeting*, S.10.

9

Mycoinsecticides

1   Introduction

Many insect pests are susceptible to infection by naturally occurring insect
pathogenic fungi. Several fungi have been studied as potential mycoinsecticides.
These fungi are very specific to insects, often to particular species, and do not
infect animals or plants. Fungi provide a needed control of insects with sucking
mouth parts. While, bacteria and viruses must be ingested to cause disease, fungi
can cause infection by penetrating the outer structure of insects. Fungi provide
the only satisfactory microbial means of biocontrol of plant sucking insects such
as aphids and white flies that are not susceptible to bacteria and viruses. These
are living, infectious microbial agents that have contact activity like many
chemical insecticides. However, they are slow acting and take about 3-7 days to
kill their insect hosts; in this regard their use is analogous to insect growth
regulators.

1.1   Mycoinsecticides for Agricultural and Forest insect Pests

McCoy et al. (1988) listed high virulence, broad host range, amenability to mass
production and formulation, storage and product stability as factors that fungi

must possess to be potential mycoinsecticides. Several fungi have been discovered that possess these attributes and have been developed asmycoinsecticides. The well-studied insect pathogenic fungi include *Beauveria bassiana* for white flies, locusts and beetles, *Metarhizium anisopliae* and *M. flavoviride* for locusts and *Verticillium lecanii* for control of aphids. Other possible fungal candidates include *B. brongniartii, Hirsutella thompsonii, Paecilomyces fumosoroseus, P. farinosus, Nomuraea rileyi* and *Aschersonia aleyrodis.*

Deuteromycetes (imperfect fungi) such as *Metarhizium anisopliae* and *Beauveria bassiana* are the causative agents of green and white muscardine diseases, respectively. *B. bassiana* is a common soil-borne saprophyte fungus that occurs worldwide. It attacks a wide range of both immature and adult insects. The extensive list of hosts includes such important pests as white flies, aphids, grasshoppers, termites, Colorado potato beetle, Mexican bean beetle, boll weevil, cereal leaf beetle, bark beetles, lugus bugs, chinch bug, fire ants, European corn borer, codling moth and Douglas fir tussock moth. There are many different strains of the fungus that exhibit considerable variation in virulence, pathogenicity and host range.

Metarhizium spp. is a natural enemy of corn root worm, white grubs and some root weevils. It also has a very broad host range and is extensively used in Brazil against spittle bugs in sugar cane and alfalfa. There are several other species of insect-pathogenic fungi that have been tested as microbial insecticides. *Verticillium lecanii* is used in Europe against greenhouse whitefly and thrips and aphids, especially in greenhouse crops. *Hirsutella thompsonii* infects mites and *Entomophthora muscae* infects flies.

Entomophaga maimaiga, a native of Japan, is an important fungal pathogen of gypsy moth (*Lymantria dispar*) larvae. Under severe disease situations, this fungus can reduce gypsy moth populations as much as 85%. The fungus survives both the winter and the absence of suitable hosts as a thick-walled 'resting spore' in the soil and on tree bark. The resting spores have been used effectively to spread the fungus to gypsy moth infested sites. In summer, resting spores germinate and produce sticky spores at the end of a stalk that grows just above the soil surface. Gypsy moth caterpillars come into contact with these spores as they search for suitable leaves to feed on. The fungus digests its way through the exoskeleton of the caterpillar and grows inside its body. Infected caterpillars may die within one week. When young caterpillars are affected early in the summer, the fungus will produce a second type of spore called conidia. These microscopic spores are spread by the wind and can infect other caterpillars. The cycle of conidia production and infection may occur four to nine times during the summer (Hajek et al., 1996). Like most fungi, its spores need moisture and high humidity to germinate. Temperatures of 50-80° F enhance fungal growth.

1.2   Mycoinsecticides for Mosquito Control

Fungi such as *Culicinomyces clavosporus, Tolypocladium cylindrosporum* and *Lagenidium giganteum* are pathogens of mosquito larvae (Gardner and McCoy, 1992). *L. giganteum* consist of asexual zoospores that are infectious to mosquito larvae. Application of *C. clavosporus* conidia or hyphae to mosquito-infested natural and artificial ponds in the United States and Australia has yielded 86-100% control of *Culex* spp., *Aedes* spp., and *Anopheles* spp. The fungus can be mass-produced as conidia or mycelia in surface or submerged culture and can recycle in the mosquito population. Similarly, both conidia and blastospores of *T. cylindrosporum* are infective to mosquito larvae and it has potential as a mosquito mycoinsecticide.

2   Mode of Action

Insect pathogenic fungi have 'contact activity', much like chemical insecticides. Their ability to invade actively the external skeleton or cuticle of insects, makes them pathogen of choice for sucking pests, whose mouth parts may preclude uptake of other pathogens such as bacteria and viruses (Figure 9.1). The site of invasion by fungi is often between the mouth parts at inter-segmental folds or through spiracles, where locally high humidity promotes germination and the cuticle is soft and more easily penetrated (Charnley et al., 1997). The host insects most commonly affected include, aphids, white flies, locusts and grasshoppers, lepidopterous larvae, ants, termites and ground beetles in foliar and soil habitats (Burges, 1998).

M. *anisopliae* and *B. bassiana* have hydrophobic spores which appear to bind to insect cuticle, as it is picked up by the insect from the environment (soil or plant surface) during feeding or movement. Once on the cuticle, the

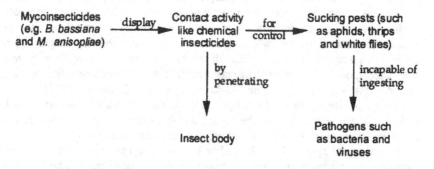

Figure 9.1   Contact activity of mycoinsecticides.

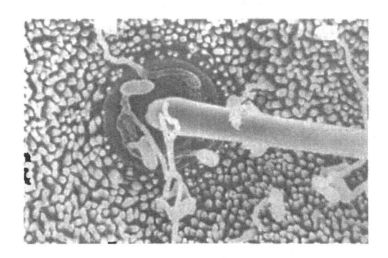

Figure 9.2   Scanning electron micrograph of *Metarhizium anisopliae*
fungal spores germinating on the surface of cuticle of the
tobacco budworm (St Leger et al., 1988)).

spore responds to biochemical cues present (chemotaxis) in the waxy epicuticle
and germinates within 8-16 hrs. (Figure 9.2). Soon the fungus stops growing
horizontally on the surface of the cuticle and initiate penetration, using a
combination of mechanical pressure and a mixture of cuticle degrading enzymes
(lipases, proteases and chitinases), which attack and dissolve the cuticle. Once
the fungus breaks through the cuticle and underlying epidermals, it tends to
invade in haemocoel (body cavity) of the insect and proliferate in the
haemolymph. The insect's defense system in the homocoel employs
phagocytosis and the secretion of antagonistic compounds namely quinones and
melamines. However, once inside the insect, entomopathogenic duteromycetes
such as *Beauverea* produces a toxin called Beauvericin that weakens the hosts
immune system. Usually, within 24 hrs. of germination, the fungus rapidly
proliferates through the insect. Growth can be in the form of mycelium or yeast-
like blastospores. The infected insects stop feeding and become lethargic. They
may die relatively rapidly within 2-7 days (Figure 9.3) (Jaronski, 1997).
        The life cycle of the fungus is completed when it sporulates on the
cadaver of the host. Under the right conditions, particularly higher relative
humidity, the fungus will break out through the body wall of the insect
producing aerial spores. High humidity is critical to spore germination, fungal
survivorship and transmission from host to host. The dead insect's body may be
firm and "cheese-like" or an empty shell, often but not always with green, red or

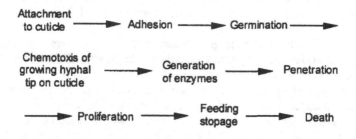

Figure 9.3   Mode of action of the insect pathogenic fungus.

brown fungal growth, either enveloping the body or emerging from joints and body segments. These external hyphae produce conidia that ripen and are released into the environment completing the cycle. This may allow horizontal or vertical transmission of the disease within the insect population.

Death of an insect from fungal infection is probably the result of starvation or physiological/biochemical disruption brought about by the fungus. For example, *B. bassiana* kill their host by depleting the insect's energy reserves (physiological starvation). It also causes a decline in fecundity in Colorado beetles that survive infection. Similarly, *M. flavoviride* causes reduction in feeding and flight of desert locusts. Another possible mechanism that may contribute to the demise of the insect is the occurrence of secondary metabolites that are insecticidal. For example, *M. anisopliae* produce a variety of toxic metabolites that act as neurotoxins (e.g. the destruxins) or general metabolite disruptors (the viridoxins). It is reported that some insects infected with entomopathogenic fungi crown to the top of the plant to die (summit disease).

3   Production of Mycoinsecticides

Production of entomopathogenic fungi on solid substrates give rise to conidium that are the natural form of spored Hyphomycetes dispersed by rain and/or air currents. This form is both stable and convenient for dispersion. On the other hand, submerged, liquid culture fermentation produces blastospores, a yeast-like phase of vegetative growth, that are both infectious and germinate faster than aerial conidium. The blastospores are environmentally sensitive particularly to desiccation. This makes their shelf life limited to a few months under refrigerated conditions. Various Deuteromycetes fungi species such as *B. bassiana, M. anisopliae, M. flavoviride, P. fumosoroseus* and *P. farinosus* are mass produced through submerged, liquid culture fermentation. Mass production of several entomopathogenic fungi has been reviewed (Bartlett and Jeronski, 1988).

For solid substrate production, nutrient rich cereal grains provide maximum surface area on which conidia can be produced. Rice is widely used, but millet is also found good as nutrient substrate for harvest convenience. The grain is broken to maximum surface area, soaked or boiled to achieve the requisite moisture content and sterilized. Heat provides increased availability of nutrients to the fungus. The fungal bio-mass builds up in the broth, on the cereal nutrients. Addition of calcium carbonate or calcium sulfate can increase the pH and prevent grains sticking to each other. The grains can be supplemented with nutrients or nutrient-socked porous mineral granules can replace it. Mineral granules such as pumice, exfoliated vermiculite and clay, can be retained as granular formulation or disintegrated to facilitate spraying (Guillon, 1997). Heat-sealable spawn-growing mushroom bags allow efficient aeration during incubation and grain drying under near 100% relative humidity. The conidia are well preserved during the sporulation period until harvest. Harvesting is done by extraction of dry spores by an air stream in a fluid-bed device forming a virtually pure spore powder. Expensive drying and harvesting can be avoided by marketing the growing bags as end-use containers, even before sporulation has peaked, providing extended shelf life by taking advantage of survival characteristic of conidia (Burges, 1998).

Mycotech Corporation has developed a commercial–scale production facility for aerial conidia of *B. bassiana* (Wraight and Carruthers, 1999*)*. Blastospores produced in a liquid medium in conventional fermenter, are incorporated into a proprietary solid medium that is loaded into trays in large chambers with forced aeration in a controlled environment. Profuse sporulation is initiated throughout the substrate within a few days. After the culture matures, the spores are dried within the chamber at a controlled rate to approximately 5% moisture content and then harvested directly from the chamber. The extracted product is a nearly pure conidial powder containing $1.2–1.8 \times 10^{11}$ conidia/g. This production technology is adaptable to other fungal pathogens that can be mass-produced on solid substrates, especially *Metarhizium* and *Paecilomyces* spp.

Culture conditions can influence the characteristics of fungal spores and can be manipulated to increase mycoinsecticidal efficiency. Blastospores of *B. bassiana* from nitrogen limited culture, were found more virulent (lower lethal time, LT_{50}) towards rice green leafhopper, than blastospores from carbon limited cultures (Lane et al., 1991).

4 Formulation of Mycoinsecticides

The mycoinsecticides can be formulated in a manner similar to conventional pesticides, i.e. as foliar sprays, soil drenches, granules and baits. Nevertheless, several unique features of microbial pest control agents have to be considered in the formulation. The spores of entomopathogenic fungi are the agents of fungal dissemination and infection in nature that are sensitive to desiccation. These fungal spores have to be kept alive in a dormant state for a reasonable time. Also that under favorable environmental conditions, these fungi have recycling potential after initial application.

4.1 Dormancy of Fungal spores

There are three essential components of conidial germination (a) nutritional source, (b) oxygen and (c) water. Elimination of any one component would prevent germination, yet killing the conidium would be avoided. Residual nutrients in a typical mass production harvest are sufficient to initiate germination of spores under sufficient moisture and oxygen. Excluding of oxygen through vacuum packaging or replacing the container headspace with N_2 or CO_2 adversely effects shelf life. On the other hand, it has been shown that exclusion of water or its reduction below certain level can prevent germination in case of *B. bassiana*, *V. lecanii* and *M. flavoviride*. However, inherent physiological characteristics of fungal species or even an isolate within a species, gave distinct survival patterns.

4.2 Hydrophobicity of Fungal spores

The fungal conidia of important fungi such as *Beauveria* spp., *Metarhizium* spp. and *Paecilomyces* spp. are extremely hydrophobic. Therefore, oils are highly compatible with lipophilic conidia as well as with the insect cuticle-leaf cuticle target system. Accordingly, formulations as oil flowable (OF) or emulsifiable suspension (ES) seem to work well. Oil carrier seems to enhance conidial contact with the insect cuticle, thus enhancing the efficacy of entomogenous fungi. Petroleum based oil carriers stabilize fungal conidia providing good shelf life even at elevated temperatures, as compared to short shelf life in plant derived oils. In the latter, presence of short chain fatty acids is known to be toxic to conidia. However, there are inconsistencies in the literature about the stability and the use of paraffinic and vegetable oils (Wraight and Carruthers, 1999). Oil formulations are generally designed for ultra low volume (UL) and undiluted applications. However, often water-diluted applications are required on vegetable crops and greenhouse applications. Phytotoxicity could also be a problem with application containing more than 1% (v/v) oil. This situation requires the use of emulsifiers, which can be toxic to conidia. Hence, proper screening of emulsifier for conidial longevity is important.

   B. bassiana hyphomycete conidia formulation Naturalis™ is based on a mixture of vegetable oils and vegetable proteins and carbohydrates. In a patent this combination is claimed to enhance contact between the fungus and the insect host by acting as an arrestant and feeding stimulant (Wright and Chandler, 1995, 1996).

   Schwarz (1995) reported a new patented granular formulation of *M. anisopliae*. The granules are produced by growing fungal biomass in liquid fermentation processed as granules under controlled conditions. The fungal biomass is separated from the nutrient broth through centrifugation and pelleted by passing through a rotating screen. The fungal pellets are then dried by slowly withdrawing water in a fluidized-bed dryer. Gentle and controlled removal of water induces the cells to enter a 'resting state'. The resultant granules are vacuum-sealed in plastic to retain viability and purity. The product shows acceptable storage stability (no appreciable loss of viability for at least 12 months) under refrigeration. The formulation consists solely of fungal mycelium and contains no extra carbon source that might stimulate the growth of soil microflora antagonistic to *M. anisopliae* after application. However, It was reported that some virulent *M. anisopliae* isolates could not be formulated as granules.

   The various salient features and requisites of mycoinsecticide formulations are given in Figure 9.4.

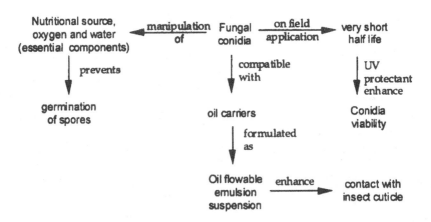

Figure 9.4   Salient features of mycoinsecticides formulation.

4.3   Post-Application Longevity of Fungal Spores

The half-life of a population of conidia directly exposed to full sunlight is a matter of hours. In case the target insect inhabits the undersides of leaves, *B. bassiana* GHA conidial half-life can be extended to several days. *Numuraea rileyii* conidia on bean and cabbage were found to have a half-life of 3.6 hours on a sunny day, but when sunlight was physically filtered, it was extended to over 40 hrs. (Fragues et al., 1988). Several UV protectants can be used as an adjuvant in the formulation of a mycoinsecticide to enhance the conidial viability.

4.4   Commercial Mycoinsecticides

Several entomopathogenic fungi are now available as commetrcial mycoinsecticides (Table 9.1). *Beauveria bassiana* can be mass-produced by a fermentation process and formulated to enable the fungus to withstand UV light and temperature and humidity extremes commonly encountered in the field. Mycotech Corporation has products based on GHA strain of *B. bassiana* (Mycotrol® and BotaniGard®) as wettable powder and oil-based formulations for control of white flies and other insects. (Jaronski, 1997). Troy Bioscience has registered a mycoinsecticide product Naturalis® based on strain JW-1 of *B. bassiana*, for control of white flies and other insects (Wright and Chandler, 1996).

BioBlast® is a formulation of *Metarhizium anisopliae* for termite control by EcoScience Corporation. In another formulation, BioPath®, this fungus has been used in bait stations for control of cockroaches, although it is no longer being manufactured (Miller, 1995). Bayer obtained registration of a pelleted dry mycelium formulation of *M. anisopliae* as BIO 1020® for control of black vine weevil (*Otiorhynchus sulcatus*), a major problem on several ornamental crops in glass houses and nursery stock.

Among the principal fungal candidates studied as insect vector control agents include *Lagenidium giganteum*, *Culicinomyces clavosporus* and species of the genus *Coelomomyces* (Federici, 1995). One fungal mosquito larvicide based on a water-borne fungus *Lagenidium giganteum* (Laginex®), has been commercialized by AgraQuest, a California based company. The fungal zoospores infect larvae of all species of mosquitoes but have been claimed to be especially effective against *Culex* spp.

Table 9.1   Commercial mycoinsecticidal products.

Trade Name	Organism	Target	Producer
Mycotrol (For field crops) BotaniGard (For green house)	*Beauveria* *bassiana* GHA	Whitefly, aphids, thrips, mealy bugs, leaf hoppers and leaf feeding insects	Mycotech Corp., Butte, MT, U.S.A.
ESC 170 GH	*B. bassiana* ESC 170		EcoScience, Worcester, MA, U.S.A.
Naturalis-O	*B. bassiana* JW-1	White flies and other insects in field and green house	Troy Bioscience, Phoenix, AZ, U.S.A.
	Metarhizium *anisopliae*	Sugarcane pests and *Aenalamia varia*	Mycotech Corp., Butte, MT, U.S.A.
BioPath BioBlast	*Metarhizium* *anisopliae* ESF 1	German cockroach Termites	EcoScience, Worcester, MA, U.S.A.
BIO 1020	*Metarhizium* *anisopliae*	Black vine weevil (*Otiorhynchus sulcatus*), citrus root weevil and termites	Bayer AG Germany
PFR97	*Paecilomyces* *fumosoroseus* Apopka 97	Whiteflies, spider mites, aphids and diamondback moth in ornamentals and vegetables	Thermo Trilogy, Columbia, MD, U.S.A.
Vertalec	*Verticillium* *lecanii*	Whitefly, aphid	Tate and Lyle, UK
Laginex	*Lagenidium* *giganteum*	Mosquito larvae	AgraQuest, Inc., Davies, CA, U.S.A.

5   Concluding Remarks

An ideal biocontrol system would be an inexpensive inoculant that could be applied once or infrequently and sustain itself in the agro-ecosystem. It would involve a set of technologies to create a fungal product that can be thought of and used as a broad spectrum insecticide, having the performance and handling characteristics of classical chemical products, while retaining the environmentally favorable characteristics of a product based on a naturally occurring microorganism. This in turn, would require addressing of such key issues as manufacturing, shelf life under unfavorable conditions, ease of application and efficacy (Miller, 1995). Increase in speed of kill and reduction in moisture threshold for spore germination through genetic and physiological engineering of fungal pathogen would need to be investigated further. Studies of synergism between fungal pathogens and low doses of chemical insecticides also indicate considerable potential for integrated control applications.

References

1.   Bartlett, M. C. and S. T. Jeronski, 1988. "Mass production of entomogenous fungi for biological control of insects", in *Fungi in Biological Control Systems*, M. N. Burge, ed., Manchester University Press, Menchester, pp. 61-85.

2.   Burges, H. D., 1998. "Formulation of mycoinsecticides", in *Formulation of Microbial Biopesticides: Beneficial Microorganisms, Nematodes and Seed Treatments*, H. D. Burges, ed., Kluwer, Dordrecht, The Netherlands, pp. 131-183.

3.   Charnley, A. K., B. Cobb and J. M. Clarkson, 1997. "Towards the improvement of fungal insecticides", in *Microbial Insecticides: Novelty or Necessity?*, Symposium proceedings No. 68, British Crop Protection Council, Farnham, pp. 115-126.

4.   Fargues, F., M. Rougier, R. Goujet and B. Itier, 1988. "Effect of sunlight on field persistence of conidia of the entomopathogenic hyphomycete *Namuraea rileyi*", *Entomophaga*, 83(3), 357-370.

5.   Federici, B. A., 1995. "The future of microbial pesticides as vector control agents", *Jour. Amer. Mosq. Con. Assn.*, 11(Part 2), 260-268.

6.   Gardner, W. A. and C. W. McCoy, 1992. "Insecticides and herbicides", in *Biotechnology of Filamentous Fungi: Technology and Products*, D.B. Finkelstein and C. Ball, eds., Butterworth-Heinemann, Stoneham, MA, pp. 335-359.

7.   Guillon, M., 1997. "Production of biopesticides: scale up and quality assurance", in *Microbial Insecticides: Novelty or Necessity?*, Symposium Proceedings No. 68, British Crop Protection Council, Farnham, pp. 151-162.

8.   Hajek, A. E., L. Butler, S. R. A. Walsh, J. C. Silver, F. P. Hain, F. L. Hastings, T. M. Odell and D. R. Smitley, 1996. "Host range of the gypsy moth (Lepidoptera: Lymantriidae) pathogen *Entomophaga maimaiga* (Zygomycetes:

Entomopathorales) in the field versus laboratory", *Environ. Entomol.*, 25(4), 709-721.

9.   Jaronski, S. T., 1997. "New paradigms in formulating mycoinsecticides", in *Pesticide Formulations and Application Systems, Vol. 17*, G.R. Goss, M. .J. Hopkinson and H. M. Collins, eds., American Society for Testing of Materials, pp. 99-112.

10.   Lane, B. S., A. P. Trinci and A. T. Gillespie, 1991. "Influence of cultural conditions on the virulence of conidia and blastospores of *Beauveria bassiana* to the green leafhopper, *Nephotettix virescens*, *Mycol. Res.*, 95, 829-833.

11.   McCoy, C. W., R. A. Samson and D. G. Boucias, 1988. in *CRC Handbook of Natural Pesticides, Vol. V. Microbial Insecticides Part A; Entomogenous Protozoa and Fungi*, C.M. Ignoffo, ed., CRC Press, Boca Raton, FL, pp 151-236.

12.   Miller, D. W., 1995. "Commercial development of entomopathogenic fungi", in *Biorational Pest Control Agents: Formulation and Delivery*, F.R. Hall and J.W. Barry, eds., American Chemical Society, Washington, DC, pp.213-220.

13.   Schwarz, M. R., 1995. "*Metarhizium anisopliae* for soil pest control", in *Biorational Pest Control Agents: Formulation and Delivery*, F.R. Hall and J.W. Barry, eds., American Chemical Society, Washington DC, pp. 183-196.

14.   St Leger, R. J., P. K. Durrands, A. K. Charnley and R. M. Cooper, 1988. "Role of extracellular chymoelastase in the virulence of *Metarhizium anisopliae* for *Manduca sexta*", *J. Invertebr. Pathol.*, 52, 285-293.

15.   Wraight, S. P., and R. I. Carruthers, 1999. "Production, delivery and use of mycoinsecticides for control of insect pests on field crops", in *Biopesticides: Use and Delivery*, F.R. Hall and J.J. Menn, eds., Humana Press, Totowa, NJ, pp. 233-269.

16.   Wright, J. E., and L. D. Chandler, 1996. "Naturalis™, a biopesticide (*Beauveria bassiana* JW-1) for control of economic insects in field crops, vegetables, ornamentals and greenhouses, with emphasis on control of *Bemisia*", *Proc. XX Int'l. Cong. Entomol., Firenze, Italy*, p. 698.

17.   Wright, J. E., and L.D. Chandler, 1995. "Biopesticide composition and process for controlling insect pests", U.S. Patent 5413784.

Part III

Integrated Use and Commercialization of Biopesticides

10

Integrated Use of Biopesticides
and Synthetic Chemical Pesticides

1 Introduction

Adverse environmental impact, increasing instances of pest resistance and public concern about food safety are among the major driving forces behind the multifaceted efforts towards the development of biopesticides. These include bioinsecticides for control of insect pests of agriculture and public health concern, biofungicides for the control of plant diseases on agronomic and horticultural crops, and bioherbicides for control of weeds. Several scientific groups searching for biological means for control have addressed a number of significant pests, fungal diseases and weeds. The primary considerations in developing such products have been a) cost effectiveness, b) user friendliness and c) adaptation with existing pest control programmes and d) their consistency and efficacy.

Since the beginning of pesticide use for agricultural purposes, combinations of two or more chemicals for sequential or simultaneous application on a single crop have been routinely used. Often, these combinations produced interactions with either increased or decreased efficacy, termed as synergistic or antagonistic effects. Synergistic approach has been widely recommended as a resistance management strategy. Synergistic action occurs, in broad terms, where the activity of a combination is much greater than that expected from their individual effects. A combination of biopesticide and synthetic chemical pesticide can also result in reduction in quantity of the highly toxic chemical, and/or reduction in the cost of pest control application.

2   Integrated Use of Biofungicides

Biological control systems of crops against phytopathogenic fungi are essentially using one organism (the biocontrol agent) to control another organism (the pathogen). Biological control systems can be developed as an alternative to fungicides or for integrated use in combination with fungicides. Apparently, the integrated approach is more reliable as biological control systems on their own are subjected to widely varying environmental influences, which in turn affect their efficacy. Indeed, biocontrol systems have been far more successful with protected crops and post-harvest control where there is a degree of control on environmental parameters.

A desirable objective of employing a mixture of two or more pesticides is to have synergistic action. Advantages of synergistic combinations might include utilization of a lower dose rate due to increased efficiency, a broader spectrum of activity and reduced risk of resistance development. Fungicides are widely used as combination of two or more active ingredients. A formulated fungicide mixture often contains the 'at risk' fungicide in combination with a partner with a different mode of action. This is one of the practiced methods of delaying the onset of resistance, which include, use of fungicide combination formulations and substitution of a dissimilar material with distinctly different mode of action for every other application. In fungi-toxic combination products, to alleviate the risk of fungal resistance due to single site fungicides, mixture partners are often non-systemic contact fungicides that operate via a multi-site mode of action.

2.1   Synergistic Combination of Bacterial Antagonists

Similar to fungi-toxic combination products, integration of more than one bacterial antagonists may provide more effective biocontrol than the use of individual antagonists on their own. Enhanced biological control of plant pathogens using two or more antagonists has been experimentally demonstrated for a number of pathosystems. Specific combinations of different strains of fluorescent Pseudomonads were shown to suppress take-all disease of wheat infested with *G. graminis* ver. *Tritici*, whereas strains used individually did not give comparable control (Pierson and Weller, 1994). Similarly, it was demonstrated that postharvest decay of apples caused by *Botrytis cinerea* and

Penicillium expansum was controlled by a combination of *Acremonium breve* and *Pseudomonas* spp. (Janisiewicz, 1988).

A number of *Bacillus* spp. have been tested and found to be suitable for development as biofungicides against phytopathogenic fungi. These include *Bacillus brevis, B. polymyxa, B. cereus, B. licheniformis* and *B. subtilis*. Some of these are antagonistic to both mycelial growth and conidial germination, whilst others are antagonistic to only one or the other of these development stages. A *Bacillus* antagonist to one stage of development of *Botrytis* (e.g. conidial germination) might be combined with an antagonist to another stage of development (e.g. mycelial growth) to provide more effective control of grey mold. However, for integrated disease control using a combination of *Bacillus* antagonists, it is important to know the modes and mechanisms of biocontrol, in order that additive effects may be achieved. For example, some combinations such as, *B. licheniformis* and *B. polymyxa* showed inhibition of each other *in vitro* studies, and should be avoided as a combination (Seddon et al., 1996).

2.2 Synergistic Combination of Fungal and Bacterial Antagonists

A unique combination of fungal antagonist *Gliocladium virens* and a bacterial antagonist *Burkholderia cepacia* was reported to be highly effective in controlling several serious soil-borne diseases of corn, tomato, and pepper in greenhouse and field tests (Lumsden, 1999; Mao et al., 1998a)). The diseases (and respective pathogens) controlled include damping-off, root rot, and stalk rot of corn in the field; seed rot and damping-off (caused by *Pythium ultimum* and *Rhizoctonia solani*) and Southern blight (*Sclerotium rolfsii*) on both tomato and pepper plants, as well as *Fusarium* wilt (*Fusarium oxysporum*) of tomato and *Phytophthora* blight (*Phytophthora capsici)* of pepper. The biocontrol treatments were applied as seed treatments and root drenches. The biocontrol treatments significantly reduced disease, as measured by plant stand, disease severity, plant fresh weight, and fruit yield, both when applied alone or in combination. The combination treatment resulted in improvements by reduced disease severity and enhanced fresh weight for pepper and fruit yield for tomato in the field, compared to that of either antagonist when applied alone.

Similar use of naturally occurring, beneficial fungi and bacteria has been made to protect corn seed from disease caused by the fungal plant pathogens *Pythium* and *Fusarium* (Mao et al., 1998b). In one field study, only about half the seeds sprouted and grew to mature plants in plots harboring both fungi. But in plots where seeds were coated with a combination of beneficial fungi and bacteria, more than 80 percent became full-grown plants. This bettered or equaled the performance of seed protected with coatings of any of several fungicides. Protection by the good microbes continues after the seedling stage. Mature plants in the biocontrol plots had about 25 to 40 percent less damage from root and stalk rot diseases, compared to plants grown from untreated seed.

Enlisting multiple species of beneficial microorganisms is a new form of biological control. In the past, the approach has been to use one biocontrol agent against one plant pathogen. But this typically does not guard against other disease-causing organisms that may also be in the soil. Advancements and improvements have also been made in the area of shelf-stable formulations and effective delivery systems for this combination biocontrol treatment (Lumsden, 1999).

2.3   Synergistic Interactions with Fungicides

Several reports have appeared stating the synergistic phenomenon involved in the integrated control using fungicides and fungal antagonists may be more efficient and longer lasting than the control achieved through fungal antagonists or fungicides alone. However for successful integration of biological and chemical control of plant pathogens, the system must be compatible. In several cases, integration of biocontrol agents with fungicides has been shown to be beneficial reducing the use of fungicides in the agricultural environment (Table 10.1).

Use of *Trichoderma harzianum* T 39 (Trichodex) alone or in combination with fungicides has been reported for control of *Botrytis* diseases in green house crops and in vineyards (Elad et al., 1994). Alternation with fungicides resulted in as good disease suppression as that achieved by the fungicide alone and more consistent than that by biocontrol alone. This approach has been tested on tomato and strawberry for control of *B. cinerea* with minimum number of chemical sprays. Similar observations were made in the control of *Botrytis* bunch rot of grape using *T. harzianum* and half rates of dicarboximide fungicide iprodione which provided extremely effective control (Harman et al., 1996).

Significant synergistic interactions with fungitoxic compounds (which are as much as 50 times as active as individual enzyme) have been reported (Lorito et al., 1994). For example, a fungi-toxic sterol, miconazole was needed in 300 ppb for achieving 50% control (ED_{50}) of *B. cinerea*. The addition of 10 ppm endochitinase enzyme from *T. harzianum* reduced the amount needed of miconazole to only 70 ppb. The mode of synergistic action postulated is that chitinolytic enzymes weaken the cell wall of the target pathogen and consequently facilitate uptake of fungitoxic compound.

Integrated treatment of cottonseed with *Gliocladium virens* at reduced levels of metalaxyl fungicide is reported to result in a synergistic action (Howell, 1991). The suppression of *Pythium ultimum* damping-off disease of

Table 10.1 Synergistic fungal antagonists and chemical fungicide combinations

Disease (Target Pathogen)	Fungal Antagonist	Chemical Fungicide	Application Mode
Botrytis diseases in green house and in vine yards, such as, grey mold in tomato and strawberry (B. cinerea)	Trichoderma harzianum T39 (Trichodex)	Different fungicides	Alternated
Crown and root rot of apple trees (Phytophthora cactorum)	Enterobacter aerogenes B8	Metalaxyl	Alternated
Botrytis diseases (B. cinerea)	Bacillus antagonists such as B. brevis	Dicarboximides such as vinclozolin and iprodione	
Damping-off of cotton seedling	Gliocladium virens	Metalaxyl	Combination
Fusarium root rot disease and pre emergence damping off of lentil (Fusarium avanaceum)	Bacillus subtilis	Vitaflo-280 (A combination formulation of carboxin and thiram	Combination

cotton seedling achieved was found to be equal to full strength fungicide treatment.

Crown and root rot of fruit trees is primarily caused by *Phytophthora cactorum*. The application of *Enterobacter aerogenes* (B 8 strain) as a soil and trunk drench reduces the percentage of apple trees with crown rot infection. Alternated application of *E. aerogenes* and metalaxyl is reported to significantly reduce infection of *P. cactorum* and increase in fruit yield (Utkhede and Smith, 1993).

Dicarboximides (vinclozolin and iprodione) reportedly show nil or low-level effects against *Bacillus* antagonists and are best suited for integration with them (Seddon et. al., 1996). It is of interest to note that gramicidins, produced by *Bacillus* spp. such as *B. brevis* and the dicarboximides, both act on membranes albeit by different suggested mechanisms of action. It has been proposed that dicarboximides act on membranes by free radical reactions with primary effects on lipid peroxidation. On the other hand, gramicidins are thought to act by disrupting the integrity of membranes by interaction with the phospholipids. It is therefore, likely that the combined action of gramicidins and dicarboximides is greater than these two components acting alone in that gramicidins might provide easier access for the dicarboximide to the membrane lipids, and thereby more effective peroxidation.

Bacillus subtilis strain GB03 suppresses pathogenic *Fusarium* spp. and *R. solani*. It is reported that *B. subtilis* GB03 in combination with standard chemical fungicides gave improved results from *R. solani* inoculated plots. The success of *B. subtilis* GB03 in the cotton market is attributed due to its integration with standard chemical fungicides. It is postulated that *B. subtilis* GB03 supplements standard chemical fungicides through an early synergy, expands the activity spectrum and provides long-term activity (Brannen and Kenney, 1997). A combination product of *B. subtilis* GB03 with metalaxyl and PCNB has been commercialized under the trade name System 3 Seed Treatment by Helena Chemical Company (Memphis, TN, USA) for control of *Rhizoctonia, Phytophthora* and *Pythium* diseases.

Similarly, integrated use of *Bacillus subtilis* with Vitaflo-280 (a combination formulation of carboxin and thiram), is reported as the most effective method for reducing disease severity due to *Fusarium avenaceum*, which caused root rot and pre-emergence damping-off in lentil seedling (Hwang, 1994) (Figure 10.1). *B. subtilis* produces several anti-fungal compounds including bacilysin and fengmycin which are inhibitory to root pathogens such as *F. avenaceum*. Competition between *F. avenaceum* and *B. subtilis* for nutrients and infection sites may also play a role in biological control. In this particular case it was found that Vitaflo-280, at higher concentration (50 ml / l) was non-toxic to *B. subtilis*, whereas it was highly toxic to *F. avenaceum* at lower concentrations (1 ml/l). Thus the two systems have potential compatibility for an integrated application for disease control. In effect, dual *B. subtilis* and Vitaflo-280 treatments significantly improved the efficacy. In this integrated system, seeds were better protected and grew more rapidly in *F. avenaceum*-infested soil than if they were treated similarly with Vitaflo-280 alone.

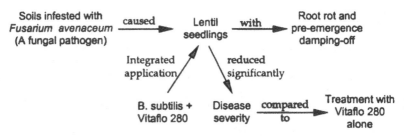

Figure10.1 Integrated treatment of *B. subtilis* and Vitaflo improved disease control as compared to treatment with Vitaflo only.

3   Integrated Use of Bioherbicides

A useful scenario would be combining one chemical or herbicide with another herbicide, in quantities less than those required when either component is used alone to provide superior and/or more economically advantageous weed control. Mechanism for this phenomenon includes combination of compounds that act at different molecular sites or that chemically block metabolic degradation of another compound. Herbicide mixtures fall into two categories (often overlapping). In the first type, each herbicide is used at a full rate and the spectrum of weeds controlled by each is mutually exclusive. In the second type, both control the same weeds.

There have been some analogous results using chemical interactions with bioherbicides that have provided beneficial, additive or synergistic enhancement of bioherbicide activity. If synergistic interactions are found those expand the bioherbicide host range, options are available to maximize the host regulation potential.

3.1   Rationale of Synergistic Combinations

Likely scenarios of synergistic/additive interactions of chemicals and bioherbicides for weed control are as follows:
a)   Weed defenses are lowered using herbicides (or any other chemical), making weeds more susceptible to pathogen attack;
b)   Bioherbicide concentration and quantity, chemical (herbicide or other) concentration, or both are reduced and
c)   Target weed range of a given pathogen is expanded by the use of a chemical synergist.

Plant (weed) defenses (physical/chemical barriers and biochemical responses) attempt to protect plants against attack from essentially all micro-organisms. Pathogens including viruses, bacteria and fungi, possess infection mechanisms that allow them to either evade or break down plant defenses, causing infection that can lead to injury or death. Herbicides, that by design cause plant injury, may be the primary choice for acting as synergists of pathogens. According to Gressel et. al. (1996) impermanent synergies such as using chemical herbicides, may have advantages, as these provide weed control when applied and the level of the organism should later dissipate, like an environmentally sound chemical herbicide. Weeds can evolve resistance to pathogens just as they have to conventional chemical herbicides. The use of chemicals as synergists or the synergistic use of two pathogens should delay or even overcome such an evolution, just as it has with conventional synergists for chemical pesticides.

3.2   Synergistic Interactions of Fungal Pathogens and Chemical Herbicides

The fungal bioherbicides require specific conditions such as a prolonged dew period or free moisture for spore germination and infection. Thus variable environmental conditions may severely hamper effective control of weeds. To overcome this limitation, integration of fungal bioherbicides with synthetic herbicidal chemicals have resulted in synergy between the two approaches. Compromising the plant in some way with a chemical agent reduces the plant's ability to mount a defence to pathogen attack, improving the efficacy of bioherbicide. Hoagland (1996) has expanded a compilation of several synergistic interactions between fungal pathogens and herbicides as given in Table 10.2.

Table 10.2   Synergistic fungal bioherbicides and chemical herbicide combination (Adapted from Hoagland, 1996).

Fungal pathogen	Herbicide/Plant Growth Regulator	Plant
Phytophthora megasperma	Glyphosate	Soybean
Colletotrichum coccodes	Thidiazuron	Velvet leaf *(Abutilon theophrasti* Medic*)*
Colletotrichum gloeosporoides	Endothall	Water milfoil *(Myriophyllum spicatum* L.*)*
Alternaria cassiae	Glyphosate	Sicklepod *(Cassie obtusifolia*)
Cercospora rodmonil	Diquat	Water hyacinth *(Eichornia crassipes* (Mart.) Solms)
Cochliobolus lunatus	Atrazine	Barnyard grass *(Echinochloa crusgalli* (L.) Beauv.)
Fusarium lateritium	Bentazone and Acifluorfen	Florida beggarweed *(Dismodium tortuosum* (Sw.))
Pythium and *Fusarium* spp.	Glyphosate	Blackbean
Fusarium solani f. sp. cucurbitae	Trifluralin	Texas gourd *(Cueurbita texana* (Scheele) Gray)
Rhizoctonia solani	Trifluralin	Bean

Caulder and Stowell (1988) of Mycogen Corporation received a U.S. patent on the use of synergistic interactions of some herbicides and four fungal pathogens. The herbicides acifluorfen and bentazon were the most effective synergists, providing increased control of several weed hosts by their respective bioherbicides, e.g. sicklepod by *Alternaria cassiae*, northern jointvetch by *Collectorichum coccodes*, hemp sesbania by *C. truncatum* and Florida beggarweed by *Fusarium lateritium*. These interactions also pointed out that there is no universal chemical synergist for all pathogens.

Sharon et. al. (1992) studied biochemical interaction of *Alternaria cassiae* and sicklepod (*Cassie obtasifolia*). The pathogen caused an elevated level of production of a flavonoid phytoalexin in sicklepod, which was found to be fungi-toxic. Treatment with glyphosate suppressed this defense response of the weed by lowering phytoalexin production. Thus glyphosate acted synergistically with this pathogen, by suppressing weed defenses. Twenty-fold less glyphosate than is normally phytotoxic, suppressed the phytoalexin production and increased the intensity of infection. (Figure 10.2).

3.3 Synergistic Interactions of Bacterial Pathogens and Chemical Herbicides

Combinations of bacteria and chemical herbicides for enhanced weed control termed X-tend bioherbicide systems by Crop Genetics International (Hanover, MD, USA) have been reported. The herbicide sulfosate (a sulphur analogue of glyphosate), causes varying degree of plant injury to a wide variety of weeds such as pigweed, barnyardgrass, yellow foxtail and johnsongrass. Bacterial

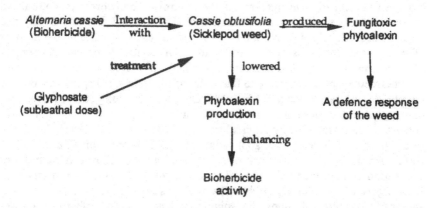

Figure 10.2 Synergy between a fungal pathogen and synthetic herbicide.

strains alone caused little or no plant injury, but sulfosate plus bacteria resulted in greater injury than from the herbicide alone (Christy et al., 1992). Similar synergistic results were obtained in field tests of two bacterial preparations with another herbicide glufosinate. The synergy between synthetic herbicides and bacterial agents could significantly reduce amounts of herbicides used to control a broad spectrum of weeds. Nevertheless, improved bacterial strains and/or formulations are needed to make successful commercialisation of such products for weed control.

4   Integrated Use of Bioinsecticides

In insect control, combination formulations have been made often using a relatively inexpensive, poorly active compound with more expensive and more active compounds. 'Salut' for example, a combination formulation of chlorpyrifos and dimethoate by BASF, gave rise to additive and synergistic action controlling a broad range of insects (Neumann et al., 1984).

Similarly, bioinsecticides such as *Bt*, in combination with many synthetic insecticides or in sequential application with synthetic insecticides has displayed synergistic activity.

4.1   *Synergistic Combinations of* Btk *with Chemical Insecticides*

4.1.1   Btk *with Endosulfan*

Integrated use of *Bacillus thuringiensis* subsp. *kurstaki (Btk)* with endosulfan has been subject of investigation for the control of bollworms of cotton. An alternated application of a *Btk* formulation (DiPel 8L) (@ 750 ml/ ha) and endosulfan (@ 2.5 l / ha) on cotton was found to result in reducing the number of sprays of synthetic insecticide from 6 to 3. The control of bollworm complex of American cotton (*Helicoverpa armigera* and *Pectinophora gossypiella*) was found effective and comparable to that with the recommended spray schedule of insecticide, increasing the yield by 52%. Over a 3 year long study, it was also observed that *Btk* alone gave an inconsistent performance, and did not prove better than recommended insecticidal treatment. (Butter et al., 1995). Similarly, in a laboratory bio-assay, an application of a *Btk* formulation (DiPel 2X) to cotton leaves at low concentration (LC_5) resulted in enhanced activity of endosulfan and reduced resistance (7-fold to 2-fold) to endosulfan in larvae of *H. armigera* (Pree and Daly (1996). The authors suggested that a mixture of *Btk* and small quantities of endosulfan might result in increased effectiveness over *Btk* alone and serve to reduce resistance to endosulfan (Figure 10.3).

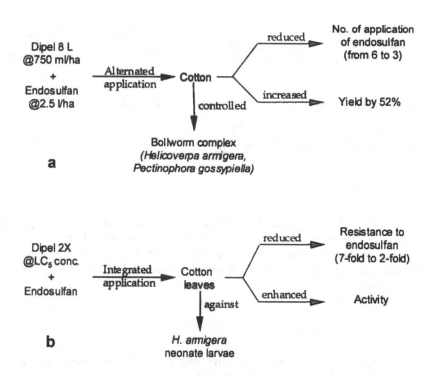

Figure 10.3 Synergy between *Bacillus thuringiensis* subsp. *kurstaki* and endosulfan, (a) alternated field applications resulted in enhanced yield of cotton and reduced number of endosulfan applications, (b) In laboratory bioassay, a tank-mix application resulted in enhanced activity and reduced resistance to endosulfan.

Cibulsky et al. (1993) have quoted the results of an evaluation of a high potency Dipel emulsifiable suspension (ES) formulation on cotton in USA and Australia. The DiPel ES was applied with a spray oil (DC-Tron) and tank mixed with endosulfan and deltamethrin (Decis) respectively for control of *H. armigera* and *H. punctigera* populations. Dipel ES in combination with endosulfan resulted in 42% less boll damage, as compared to Dipel ES by itself and Dipel ES and deltamethrin combination (Table 10.3).

Table 10.3   Evaluation of DiPel ES for *Heliothis* control of cotton
(Adapted from Cibulsky et al., 1993).

Integrated System	Rate (BIU[a]/Ha)	Spray volume (l/ha)	Boll damage (%)
DiPel	30	30	6
DiPel + Endosulfan	30	2.0 + 1.5	3.5
Dipel + Deltamethrin	30	2.0 + 1.5	7.3

(a)   BIU = Billion International Units

4.1.2   Btk *with thiodicarb*

Klein et al. (1995) investigated the role of *Btk* with ovicides in the management of the *heliothine* complex in cotton. A far greater reduction in larval population was observed with combination of *Bt* and ovicides than *Bt* alone. It was also reported that *Bt* and thiodicarb (Larvin®) provided best control of several ovicides tested.

4.1.3   Btk *with Imidacloprid*

Btk preparations in combination with imidacloprid have been claimed to enhance insecticidal activity (Schnorbach, 1995). Imidacloprid is a systemic insecticide and functions as insect nicotinergic acetylcholine receptor (nAChR). It strongly binds to acetylcholine-receptors on the nerve cells of insect blocking the binding site of acetylcholine. The resultant interruption of neural transmission paralyses the insect and it dies. Imidacloprid displays selectivity between insects and mammals. Even in common receptors that exist in insects and mammals, adequate levels of selectivity are achieved due to lesser sensitivity of receptors in mammals. Compared with rat muscles, insect nicotinic acetylcholine receptors are up to 1000 times more sensitive to the effect of imidacloprid. Thus, a combination of imidacloprid and *Btk*, with the two different modes of action, is more effective and yet safer to mammals.

4.1.4   Btk *with Mycotoxins of* Metarhizium anisopliae

A combination of *Btk* with destruxins (*Metarhizium anisopliae* mycotoxins) is reported to demonstrate synergistic interaction at low doses, in a laboratory bioassay on the 5th instar larvae of spruce budworm (*Choristoneura fumiferana*). Based on the data obtained in the bioassays involving combined agents, extrapolation of the model correlating mortality rates as a function of the various lethal doses used for each agent was plotted. It was found that a combination of two agents at their individual LD_{20} dose rates would give rise to nearly 55% mortality. It was also observed that a combination of *Btk* (at LD_{15}) and destruxins (at LD_{40}) and vice-versa concentration ratio gave rise to nearly the same mortality (~72%) indicating that both the agents have similar contribution to synergism (Brousseau et al., 1998).

4.2   Synergistic Combinations of Bti

4.2.1   Bti *with Teflubenzuron*

Chui et al. (1995) evaluated the use of *Bti* formulation (VectoBac G) with teflubenzuron for control of *Aedes aegypti* larval populations under laboratory conditions. Individually, both *Bti* and teflubenzuron were found to be highly effective in reducing *A. aegypti* larval population. However, teflubenzuron gave higher degree of residual toxicity than VectoBac G. Integration of both at their LC_{95} concentrations in different ratios revealed that a 1:9 LC_{95} concentration of *Bti* and teflubenzuron gave better control i.e. 1.35 times of LC_{95} of VectoBac and 1.18 times of LC_{95} of teflubenzuron (Figure 10.4).

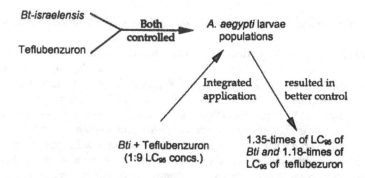

Figure 10.4   Synergistic interaction of *Bti* and teflubenzuron result in improved control of *A. aegypti*.

4.2.2   Bti *with Methoprene*

Similarly, Perich et al. (1988) reported laboratory evaluation of a formulation of *Bti* combined with methoprene against *Anopheles albimanus* and *An. stephensi*. It was found that 'Teknar', a *Bti* formulation alone as well as 'Duplex', a formulation consisting of *Bti* and methoprene, a juvenoid insect growth regulator, produced 100% mortality in both species. However an expected low mortality was found in late 4th instar larvae of both species, when treated with 'Teknar' alone. Significant mortality of late 4th instar larvae and a low adult emergence occurred with both species in water treated with 'Duplex'. This demonstrated that *Bti* can be used effectively in combination with other biocontrol agents.

4.2.3   Bti *with other Biocontrol Agents*

Combination of *Bti* with mosquito fish (*Gambusia affinis*) gave better control of *Culex tarsalis* coq. populations than when each agent was used separately. Neri-Barbosa et al. (1997) reported that *Bti* (Bactimos briquettes) combination with an insect predator species such as the back swimmer *Notonecta irrorata* uhler effectively controlled *Culex* spp. in a cost effective and environmentally safe manner. They reported efficient reduction of mosquito larvae in both Bactimos application alone and the combined Bactimos and predator application. Nevertheless, from an economical point of view, the combination approach was found better, as Bactimos had to be added less frequently to containers having notonectids in them to effect the level of control desired.

4.3   *Synergistic Combinations of Baculoviruses with Synthetic Insecticides*

A cabbage moth NPV product Mamestrin, has been found effective for control of its hosts, including *H. Zea, H. virescens, H. armigera, S.* exigua and *Diparopsis watersi*. On the other hand, Mamestrin[+], a combination product with 0.3% cypermethrin is claimed to have an enhanced host range. The added species include *S. littoralis, Plusia* sp., *Plutella xylostella, Helula undulis, Maruca testulalis* and *Phtorimaea operculella* (Cunningham, 1995).

  Similar positive interactions between several wild-type baculoviruses with photostable pyrethroids have been reported. Aspirot et al. (1987) have documented potentiation between *Mamestra brassicae* nucleopolyhedrosis (MbMNPV) and deltamethrin on *Spodoptera frugiperda, S.* exigua and *H. virescens*. Similarly, MbMNPV with fenvalerate has been shown to exhibit potentiation against diamondback moth (*Plutella xylostella*) (Bianche, 1991). A significant reduction in the LC_{50} of *Pieris brassicae* granulovirus (PbGV) was observed when combined with a low concentration of permethrin (Peters and Coaker, 1993).

Shapiro et al. (1994) observed that *Lymantria dispar* nucleo-polyhedrovirus (LdMNPV) in combination with Neem extract significantly reduced the LT_{50} for the gypsy moth compared to the LdMNPV-water treatment.

McCutchen et al. (1997) reported positive interactions (decrease in the median lethal time, LT_{50}, compared with the LT_{50} for either component alone) of recombinant *Autographa californica* nuclear polyhedrosis virus (AcAaIT) that expresses an insect selective neurotoxin (AaIT), when combined with low concentrations of several insecticides. Cypermethrin and methomyl were synergistic in combination with AcAaIT against *Heliothis virescens*. On the other hand, another recombinant virus, AcJHE.KK, that expresses a modified version of juvenile hormone esterase, showed no evidence of either synergism or antagonism in combination with cypermethrin. In another study, the results indicate that the pyrerhroids did not enhance speed of kill indiscriminately with any recombinant baculovirus.

American Cyanamid has patented combinations of recombinant insect viruses such as AcMNPV.AaIT and AcMNPV. egt- [containing a deletion in the gene encoding ecdysteroid UDP-glucosyl transferase (egt)] with several chemical and biological insecticides claiming enhanced insect control (Black et al., 1996). They observed synergism with the recombinant AaIT virus in combination with the formamidine, arylpyrrole, diacylhydrazine and cypermethrin when tested against *H. zea* larvae. In contrast, with recombinant virus AcMNPV.egt-, only diacylhydrazine significantly hastened the speed of kill. However, against *H. virescens* larvae, both the recombinant virus and cypermethrin mixtures produced synergistic interactions.

4.4 Synergistic combination of mycoinsecticides with synthetic insecticides

Entomopathogenic fungi may be used with chemical insecticides to obtain a synergistic or additive action. For example, teflubenzuron, an insecticide which selectively interrupts chitin synthesis of insects – but not fungi – synergized mortality of third-instar larvae of desert locusts with *M. flavoviride* (Joshi et al., 1992). Similarly, it is reported that the combinations of either *B. bassiana* or *M. anisopliae* (10^6 to 10^7 conidia/ml) with sublethal doses (100 ppm or greater) of imidacloprid increased synergistically mortality and mycosis of 1st instars of citrus root weevil *Diaprepes abbreviatus* (Quintela and McCoy, 1997). The larvae of citrus root weevil are of significant economic importance because of the severe root injury they cause to the citrus tree. The lethal time was also reduced significantly through the synergistic interaction of fungal/chemical treatment.

A synergistic combination of buprofezin, and *B. bassiana* has been patented which is claimed to be highly effective against a broad spectrum of different insects including scale insects, whitefly, leafhoppers and thrips (Knauf and Morales, 1999). The two components can be applied simultaneously or in succession with the advantage of long residence time and use of mycoinsecticide at a considerably low cost.

5   Concluding Remarks

It appears unlikely, that biological controls will completely substitute for chemical controls, at least in the immediate future. Therefore, the integrated use of biological control agents with existing pest control strategies within individual cropping systems may be a more viable approach. Alternating or combining biological and chemical strategies may increase efficacy of insect pest, weed and disease control and possibly prolong the efficacy of chemical pesticides by avoiding or retarding, the development of resistance in target organisms. In addition, integration of biological control strategies with crop management practices, such as the manipulation of physical chemical soil conditions to favor biological control agents would promote their efficacy and simultaneously, increase the sensitivity of the pathogen towards the agents.

Coexistence and interaction of microorganisms is the norm in nature. Due to co-operation among several populations of microorganisms, soil suppressiveness usually limits disease incidence efficiently and consistently. On the contrary, biological control based on the application of single antagonistic strain is often inconsistent. One way to improve efficacy and consistency of biological control would be to mimic the complexity of the mechanisms operating in suppressive soils, and to use several populations of antagonistic microorganisms together. With mixed biocontrol inoculum, potentially increased diversity of antibiotics being produced and nutrients being utilised could greatly increase the effectiveness of the product. It has been demonstrated, for example, that co-inoculation of fluorescent *Pseudomonas* with root colonizing fungi always resulted in greater and more consistent suppression of disease than the application of single microorganism (Leeman et al., 1996).

References

1.   Aspirot, J., G. Biache, R. Delattre and P. Ferron, 1987. "Process for the biological control of insects which destroy crops and insecticidal compositions", U.S. Patent No. 4,668,511 (May 26, 1987).
2.   Bianche, G., 1991. "Method for the biological fight against the crop ravaging insect, *Plutella xylostella*, using a nuclear polyhedrose and at least one synthetic pyrethrinoid", U.S. Patent No. 5,075,111 (December 24, 1991).

3.   Black, B. C., C. F. Kukel and M. F. Treacy, (American Cyanamid), 1996. "Mixture of genetically-modified insect viruses with chemical and biological insecticides for enhanced insect control", PCT Int. Appl. WO96-3, 048, 8 Feb.1996, *Chem. Abstr.*, 124, 310294 x.

4.   Brannen, P. M. and D. S. Kenney, 1997. "Kodiak – a successful biological-control product for suppression of soil-borne plant pathogens of cotton", *J. Ind. Microbiol. Biotechnol.*, 19, 169–171.

5.   Brousseau, C., G. Charpentier and S. Belloncik, 1998. "Effects of *Bacillus thuringiensis* and Destruxins (*Metarhizium anisopliae* mycotoxins) combination on spruce budworm (Lepidoptera: Tortricidae)", *J. Invertebr. Pathol.*, 72, 262-268.

6.   Butter, N. S., G. S. Battu, J. S. Kular, T. H. Singh and J. S. Brar, 1995. "Integrated use of *Bacillus thuringiensis Berliner* with some insecticides for the management of Bollworms on cotton", *J. Ent. Res.*, 19(3), 255-263.

7.   Christy, A. L., K. A. Herbst, S. J. Kostka, J. P. Mullen and I. S. Carlson, 1992. "Synergising weed biocontrol agents with chemical herbicides", in *Pest Control with Enhanced Environmental Safety*, S. O. Duke, J. J. Menn and J. R. Plimmer, eds., American Chemical Society, Washington DC.

8.   Chui, V. W. D., K. W. Wong and K. W. Tsoi, 1995. "Control of mosquito larvae (Diptera: Culicidae) using *Bti* and teflubezuron: Laboratory evaluation and semi-field test", *Environ. Int.*, 21(4), 433-440.

9.   Cibulsky, R. J., B. N. Devisetty, G. L. Melchior and B. E. Melin, 1993. "Formulation and application technologies for microbial pesticides: Review of progress and future trends", *J. Testing Eval.*, 21, 500-503, ref.9.

10.   Caulder, J. D. and L. Stowell (Mycogen Corporation), 1988. "Synergistic herbicidal compositions comprising *Alternaria cassiae* and chemical herbicides", U.S. Patent 4,776,873 (October 11, 1988).

11.   Cunningham, J. C., 1995. "Baculoviruses as microbial insecticides", in *Novel approaches to Integrated Pest Management*", CRC Press, pp. 261-292.

12.   De, R. K., R. G. Chaudhary and Naimuddin, 1996. "Comparative efficacy of biocontrol agents and fungicides for controlling chickpea wilt caused by *Fusarium oxysporum* f. sp. *ciceri.*", *Indian J. Agri. Sci.*, 66, 370-373.

13.   Elad, Y., D. Shtienberg and A. Niv, 1994. "*Trichoderma harzianum* T-39 integrated with fungicides: Improved biocontrol of grey mold", *Brighton Crop Protection Conf. - Pests. Dis.* Vol.3, 1109-1114.

14.   Gressel, J., Z. Amsellem, A. Warshawsky, V. Kampel and D. Michaeli, 1996. "Biocontrol of Weeds: Overcoming evolution for efficacy", *J. Environ. Sci. Health*, B 31(3), 399-405.

15.   Harman, G. E., B. Latorre, E. Agosin, R. San Martin, D. G. Riegel, P. A. Nielsen, A. Tronsmo and R. C. Pearson, 1996. "Biological and integrated control of *Botrytis* bunch rot of grape using *Trichoderma* spp.", *Biol. Control*, 7(3), 259-266.

16.   Hoagland, R. E., 1996. "Chemical interactions with bioherbicides to improve efficacy", *Weed Technol.*, 10(3), 651-674.

17.   Howell, C. R., 1991. "Biological control of *Pythium* damping-off of cotton with seed coating preparation of *Gliocladium virens*", *Phytopathol.*, 81(7), 738-741.

18.   Hwang, S.F., 1994. "Potential for integrated biological and chemical control of seedling rot and pre-emergence damping off caused by *Fusarium avenaceum* in

lentil with *Bacillus subtilis* and Vitaflo – 280", *J. Plant Disease and Protection* (Germany), 101(2), 188-199.

19. Janisiewicz, W. J., 1988. "Biocontrol of postharvest diseases of apples and antagonist mixtures", *Phytopathol.*, 78(1), 194-198.

20. Joshi, L. and A. K. Charnley, 1992. "Synergism between entomo-pathogenic fungi, *Metarhizium* spp. and the benzoylphenyl urea insecticide, teflubenzuron, against the desert locust, *Schistocerca gregaria*", in *Proc. Brighton Crop Protec. Conf. – Pests and Diseases,* Vol. 4, British Crop Protection Council, Farnham, pp. 369-374.

21. Klein, C. D., D. R. Johnson and A. M. Jordan, 1995. "The role of *Bt* plus ovicides in management of the *heliothine* complex", *Proc. Beltwide Cotton Conf., Mamphis, USA,* 2, 880-881.

22. Knauf, W. and E. Morales (Hoechst Schering), 1999. "Insect control compositions comprising entomopathogenic fungi", U.S. Patent No. 5,885,598 (March 23, 1999).

23. Leeman, M., F. M. Den Ouden, J. A. Van Pelt, C. Cornelissen, P. A. H. M. Bakker and B. Schippers, 1996. "Suppression of fusarium wilt of radish by co-inoculation of fluorescent Pseudomonas spp. and a root colonizing fungi", Eur. J. Plant Pathol., 102, 21-31.

24. Lorito, M., C. Peterbauer, C. K. Hayes and G. E. Harman, 1994. "Synergistic interaction between fungal cell wall degrading enzymes in different antifungal compounds enhance inhibition of spore germination", *Microbiology,* 140(3), 623-629.

25. Lumsden, R., 1999. "Biocontrol combination of fungal and bacterial antagonists control multiple soilborne diseases of corn, tomato, and pepper plants", *USDA's Agricultural Research Service in Beltsville,* MD, http://www.barc.usda.gov/psi/bpdl/recent.htm.

26. Mao, W., J. A. Lewis, R. D. Lumsden and K. P. Hebbar, 1998a. "Biocontrol of selected soilborne diseases of tomato and pepper plants, *Crop Protection,* 17(6), 535-542.

27. Mao, W., R. D. Lumsden, J. A. Lewis and P. K. Hebbar, 1998b. "Seed treatment using pre-infiltration and biocontrol agents to reduce damping-off of corn caused by species of *Pythium* and *Fusarium*", *Plant Diseases,* 82, 294-299.

28. McCutchen, B. F., K. Hoover, H. K. Preisler, M. D. Betana, R. Herrmann, J. L. Robertson and B. D. Hammock, 1997. "Interactions of recombinant and wild-type Baculoviruses with classical insecticides and pyrethroids-resistant tobacco budworm (Lepidoptera: Noctuidae)", *J. Econ. Entomol.* 90, 1170-1180.

29. Neri-Barbosa, J. F., H. Quiroz-Martinez, M. L. Rodrignez-Tovar, L. O. Tejada and M. H. Badii, 1997. "Use of Bactimos briquettes (*Bti* formulation) combined with the backswimmer *Notonecta Irrorata* (Hemiptera: Notonectidae) for control of mosquito larvae", *J. Amer. Mosquito Control Asso.*, 13, 87-89.

30. Neumann, U. and V. Harris, 1984. "BAS 270 001-A versatile broad spectrum dual component insecticide formulation for use on pomefruit, grapes and other crops", *British Crop Protect. Conf., Pests and Diseases,* 5A-6, p. 437.

31. Perich, M. J., J. T. Rogers, L. R. Boobar and J. H. Nelson, 1988. "Laboratory evaluation of formulations of *Bacillus thuringiensis* var. *israelensis* combined

with methoprene or a monomolecular surface film against *Anopheles albimanus* and *An. Stephensi*", *J. Amer. Mosquito Control Asso.*, 4, 198-199.

32. Peters, S. E. O. and T. H. Coaker, 1993. "The enhancement of *Pieris brassicae* (L.) granulosis virus infection by microbial and synthetic insecticides", *J. Appl. Ent.* 116, 72-79.

33. Pierson, E. A. and D. M. Weller, 1994. "Use of mixtures of fluorescent pseudomonads to suppress take-all and improve the growth of wheat, *Phytopathol.*, 84(9), 940-947.

34. Pree, D. J. and J. C. Daly, 1996. "Toxicity of mixture of *Bacillus thuringiensis* with endosulfan and other insecticides to the cotton bollworm *Helicoverpa armigera*", *Pestic, Sci.* 48(3), 199-204.

35. Quintela, E. D. and C. W. McCoy, 1997. "Pathogenicity enhancement of *Metarhizium anisopliae* and *Beauveria bassiana* in first instar of *Diaprepes abbreviatus* (Coleoptera: Curculionidae) with sublethal doses of imidacloprid", *Environ. Entomol.* 26(5), 1173-1182.

36. Schnorbach, H. J. (Bayer AG), 1995. "Insecticidal compositions comprising *Bacillus* preparations", Eur. Pat. Appl. EP 677, 247 (18 Oct., 1995), *Chem. Abstr.*, 123, 308701 P.

37. Seddon, B., S. G. Edwards and L. Rutland, 1996. "Development of *Bacillus* species as antifungal agents in crop protection", in *Modern Fungicides and Antifungal compounds, 11th International symp. Thuringia, Germany.* 1995, Intercept Ltd., Andover, Hampshire, U.K. pp. 555-560.

38. Shapiro, M., J. L. Robertson and R. E. Webb, 1994. "Effect of neem seed extract upon the gypsy moth (lepidoptera: Lymantriidae) and its nuclear polyhedrosis virus", *J. Econ. Entomol.*, 87, 356-360.

39. Sharon, A., Z. Amsellem and J. Gressel, 1992. "Glyphosate suppression of an elicited defence response: increased susceptibility of *Cassia obtusifolia* to a mycoherbicide", *Plant Physiol.*, 98, 654-659.

40. Utkhede, R. S. and E. Smith, 1993. "Long-term effects of chemical and biological treatments in crown and root rot of apple trees caused by *Phytophthora cactorum*", *Soil Bid. Biochem.* 25, 383-386.

11

Commercialization of Biopesticides

1 Introduction

Current growth in the development of biopesticides is both regulatory and technology driven. In the late 1970s, there was a growing awareness of the real and potential problems that synthetic chemical pesticides posed to human health and the environment as a whole. With this awareness, came more rigorous registration requirements that resulted into registration of fewer new active ingredients. Simultaneously, the average development cost per product got increased to $ 50 million. The problem has been compounded by the fact that pests became resistant to many pesticides, resulting in abandonment of the product or the need for higher application rates. On the other hand, improved understanding of biological processes and progress in genetic engineering techniques have given a new impetus to development of biopesticides.

In the United States, regulatory and technology driven growth of biological pest control alternatives is especially pronounced. In order to strengthen and accelerate the Environment Protection Agency's pesticide registration program, the U. S. Congress amended the federal insecticides, fungicides and rodenticides act in 1988. Because of the costs associated with the re-registration process, many pesticide companies dropped the registration of products for the small markets. This caused significant problem to small market

245

growers. These factors combined with the increasing difficulty and costs to develop and register new classes of chemical pesticides, have encouraged the search for safe alternatives for control of plant pests (Wilson and Jackson, 1997).

1.1   Fast-track registration for biologically-based technologies

In recognition of the need for pest control alternatives to chemical pesticides, the pesticide regulation authorities in many countries have the mandate to put the development of biologically based technologies for the control of plant pests on the 'fast track' for registration. The U.S. EPA has recognized that biochemical and microbial pesticides are distinguished from standard chemical pesticides and has established different data requirements as part of its registration regulations. Biopesticides currently are subject to a three-tier toxicology testing procedure and a four-tier environmental testing procedure. Tier 1 testing for EPA requires data on non-target organisms and environmental fate. A biopesticide product which satisfactorily completes both the Tier I toxicology and environmental tests is not required to go through the tests specified in subsequent tiers.

Registration authorities treat each case on its merits and according to the purpose of the program and the target pathogen, and will select the evaluation process to meet the nature of application. For instance, biofungicides for post-harvest decay control in packing houses were relieved of the need to conduct ecological testing, since the organisms are expected to be applied only in a confined environment (Hofstein and Chapple, 1999). However, should questions arise during any tier of testing, additional tests may be required. This translates into less regulatory requirements for a biologically-based system than a chemical pesticide.

Thus, as against more than 120 safety and other issues related documents to support the registration of a chemical pesticide, current US EPA requirements for a biological fungicide, for example, are for approximately a dozen studies. This difference means huge cost and time savings in favor of the biologicals. Whereas the registration cost of a chemical pesticide might be $ 10-20 million and take 5-8 years, the registration cost of a biological pesticide might be $1/10^{th}$ or less of that figure and the time period may be only 1-2 years. The high cost, long-term safety studies such as carcinogenicity, plant and animal residue and ecological fate and effects are not required for the biological control agents (Froyd, 1997). For a biopesticide product completing only Tier I testing, approximately one year of laboratory testing is required. Subsequent U.S. EPA registration generally takes approximately one year. Toxicology testing, field development trials and related costs for U.S. EPA registrations incurred for biopesticides have averaged under $500,000 for each product registration.

1.2   Genetically engineered microorganisms for pest control

In the US, regulatory authorities have taken the approach that no distinctions would be made between natural and genetically altered organisms in terms of the regulatory requirements to prove safety and efficacy. In July 1990, The US approach to regulate biopesticides was further laid down in the 'Principles for federal oversight of biotechnology: Planned introduction into the environment of organisms with modified hereditary traits'. This underlined the position that no distinctions between natural and genetically altered organisms would be made, focusing instead on characteristics of organism and environment concerned. This was based on a set of scientific principles for the evaluation of biopesticides, including the characterization of organisms, the spread into the environment, as well as effects on humans, domestic animals, wild life and non-target organisms. The first genetically engineered products approved under the new US legislation were Mycogen's biopesticides MVP® and M-Trek® (discussed in Chapter 3).

The major challenge with development and commercialization of genetically engineered biopesticides is the production of an effective product that poses little or no risk to health or the environment. These concerns include,

(a)   the release of any engineered organism into the environment raises questions of potential effects to human health and non-target organisms;

(b)   that engineered organisms will displace naturally occurring organisms from their niches, thereby causing ecological perturbations;

(c)   if an engineered organism exhibits unanticipated and deleterious environmental properties, it may be difficult to eliminate the organism from the environment and last, but not the least,

(d)   the foreign genetic material may be transferred from the released organism to other organisms with unpredicted consequences.

These issues need be satisfactorily addressed prior to the field release of genetically engineered microorganisms such as baculoviruses (Wood, 1995). There are also functional requisites touching upon mass production, effectiveness and stability for successful commercialization.

2   Commercialization Aspects

A commercial biopesticide is defined as one which works consistently in a field environment and that can be economically produced and formulated in a form that will maintain the viability of the organism through commercial distribution system. In order to be successful for commercial application, biopesticides must be efficient, dependable, cost-effective and safe for humans, the crop and the environment.

The U.S. Congress has enacted the Food Quality Protection Act (FQPA) 1996, which removed the Delaney Clause related to cancer causing pesticides that appear in processed foods. At the same time, it mandated greater margins of safety for all new pesticide tolerances. It requires that all U.S. EPA tolerances be reassessed using new standards within 10 years. All tolerances will now be based on a "reasonable certainty of no harm" and there will be a specific determination of risk to infants and children.

There is some expectation that due to new risk requirements many older products may have to be dropped. If this were to happen, the viability of biocontrols should increase, as there would be fewer cheaper chemicals to compete with. Although the investment differentials exist in the development of biologicals and chemical pesticides, the priority of the industry is market potential and profits. Therefore only those microbials showing promise of providing an economic return will be developed as biopesticides. In fact impetus for development of biopesticides may be the existence of significant pest problems that can not be solved by chemical pesticides (Gardner and McCoy, 1992). There is great optimism that biologically-based alternatives can successfully integrate with, compete with and even replace synthetic pesticides in certain applications (Froyd, 1997). Even more significant than FQPA as an agent for change is the rapid introduction and acceptance of biopesticides. Recently most of the new active ingredients registered in the United States have been biopesticides, including biochemicals and genetically modified organisms (GMOs). It is anticipated that some biopesticides will become major tools for organic farming, integrated pest management (IPM), and as replacements for products lost due to FQPA (Stewart, 1999).

2.1   Technical Constraints of Commercialization

There have been certain technical barriers that have prevented broad commercial utilization of biopesticides. Some of the key areas include fermentation, stabilization and delivery technology that needed more focus for commercialization to succeed. There is also a general problem that pertains to basic researchers, who do not understand what is needed to successfully commercialize a biological product. This eludes them in focussing their energy and resources on resolving the key technical issues (Zorner et al., 1993). Also many organisms provide excellent results under laboratory conditions but currently there is no practical way to produce an economical product from these organisms. Researchers need to be aware of industrial production and agricultural restraints. Besides, good laboratory results do not necessarily correlate to good field results. Lab tests are often conducted with biologicals in a form that does not closely resemble that produced through fermentation technology (Brannen and Kenney, 1997). It is therefore, important that the individual scientists make liaisons with people who have these skills and thus work in a multi-disciplinary environment

In 1996, the Society for Industrial Microbiology in the U.S. organized two symposia on the formulation and commercialization of biological based products for the control of plant pests. In the symposia, various technological constraints that are shared by living microbial biocontrol agents and that have hindered their commercial use for the control of insects, weeds and plant pathogens, were identified. These included (a) the lack of availability of low cost production methods, (b) the need for development of suitable microbial formulations with reasonable shelf lives, (c) the need for consistent pest control under field conditions and (d) the need to ensure that the plant pathogens that form the base of the commercial pest control products do not pose an environmental threat to other plants or segments of the eco-system in which they will be used (Wilson and Jackson, 1997).

2.2 Market Environment

Biologicals have been unable to penetrate beyond niche markets so far, and even this role is threatened as a new generation of pesticide chemistries has come on line and transgenic plants reduce the size of the sprayable pesticide market. Small niche markets tolerate the mediocre performance of biopesticides due to lack of competition and as a trade-off for their environmental and user safety. However this does not amount to a marketing opportunity but user acceptance of weakly performing product(s) within these technology-starved markets (Stowell, 1993). It is imperative that performance standards of biopesticides are raised to improve their commercialization prospects.

Various factors necessary for commercialization of biopesticides include improvement of product efficacy by technology improvements, enhancement in their price/performance characteristics and improved understanding on how to best use often in conjunction with other products. Inter–product competition between existing broad-spectrum chemical pesticides, new selective chemistries (such as Spinosad for cotton and imidacloprid for potatoes in competition with Bt), transgenics and novel technologies is intensifying for maximum market share. This in turn, would intensify the competitive pricing environment and result in rapid shift in individual product's market share. New chemistry products and transgenics would also spur the adoption of biopesticides through the expansion of the selective controls market (Shimoda, 1997).

The North American biopesticides market consists of a dual structure; large diversified companies and small-dedicated companies, which account for a significant share of the innovations. The existence of the small companies depends on their ability to protect the fruits of their research and signal their existence to the market. Indeed many small companies focus on the development of new patentable products and processes with the intention of selling the patents or their shares to larger firms. Capitalization on market opportunities and reduction of the business risk tied to a market place, favors strategic action by small companies to either a) develop critical mass, by acquisitions and mergers, b) have product(s) with huge market potential, or c) enter into strategic alliances/ virtual integration with larger companies (Shimoda, 1997).

2.3   Search for a Suitable Model and Viable Approach

The current approach to commercializing biopesticides is based on a chemical pesticide model. This paradigm emphasizes major crops and is based on cheap, stable products that are easy to scale-up and use. Apparently biological agents fit the chemical model poorly. Thus a change in approach is necessary, one based on the realities of biological systems. This approach might emulate business sectors that have successfully overcome biological limitations. A typical example is food industry. It has demonstrated that low stability products are not a fatal flaw. Similarly, the microbrewery industry has demonstrated that there is a strong market for fresh product sold without preservatives. The keys have been local small batch custom production and quick turnover leading to high quality products. The concept might be adapted to biopesticides (Gaugler, 1997).

A conference to review the progress on biological control was organized by Boyce Thompson Institute and US Department of agriculture at the Cornell University in April 1996. Subsequently, a conference on 'Biopesticides and transgenic plants - new technologies to improve efficacy, safety and profitability' was organized in January 1997 at Washington, DC. The conference deliberated on interdisciplinary analysis of commercialization challenges. This was followed by a workshop on the 'Alternative paradigms for commercializing biological control' at Rutgers University in May-June 1998. This workshop focussed attention on creating a new mindset for developing and using biological control and fostering close cooperation among industry, researchers and extension personnel interested in biocontrol. The furious activity and debate on the search of most suitable approach for commercialization of biopesticides continues in many parts of the world.

3   Commercialization of Biofungicides

Commercialization of biofungicides has received a significant boost in recent years. This has been primarily due to the impressive progress in isolation and characterization of novel strains of microorganisms that have met basic requisites for commercialization. The desirables include consistent suppression of pathogens under field conditions and easy mass production in standard fermentation facilities. The various steps involved in development of a cost-effective biofungicide are in quite rigid order and are critical for their successful commercialization. These include, isolation of active strain against specific/non-specific pests, development of bioassays, laboratory evaluations, laboratory scale-up production, formulations, field evaluations, product registration, commercial production and market acceptance of new technology (Figure 11.1).

The biocontrol agents must not only be produced in high yield but also should have high retention of cell viability with maintenance of crop compatibility and bioefficacy during several months of storage. Beyond production, a satisfactory formulation is necessary to keep the living agent stable and infectious 'on the shelf', which has to be at ambient temperature to accommodate distribution system and on farm storage conditions. An important feature for successful commercialization is the ability to devise a rapid and reliable quality control procedure that allows precise quantification of active ingredient. The biocontrol agent should also provide long-term stability under warehouse conditions, be compatible with chemical fungicides and insecticides applied to seed and be compatible with current production practices. Any deviation from these general guidelines will result in failure.

The biofungicides that have succeeded through the stage of commercialization have had a critical analysis of market needs and potential. It is envisaged that fungicides entering markets where they have the best chance of performing and where the market is receptive to using biological control methods would have better chance to succeed. The suitable products shall provide competitive performance, effective formulations and economic production. Performance testing of biocontrol agents under representative

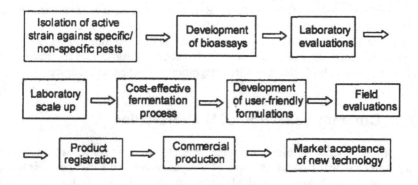

Figure 11.1   Critical steps for commercialization of biofungicides

cropping conditions, perhaps using a standard fungicide treatment as a control, will be critical to assure efficiency and dependability of new biocontrol agents and to convince growers that the new-fangled methods are worth adopting (Sutton, 1995). It is desirable to provide demonstrated financial return to producers

3.1   Biofungicide Commercialization - Case Studies

An insight into various attributes and requisites for successful commercialization can also be obtained from the case studies of various biofungicides. Harman (1996) cited his experience on commercialization of a patented biofungicide *Trichoderma harzianum* strain 1295-22 (Bio-Trek 22G) which was demonstrated to be highly effective seed protectant. He observed that legal and commercial considerations were extremely important. Particularly protection of intellectual property was essential, as without it, few companies will be interested in developing it further. Besides, adequate business, legal and financial expertise, availability of large-scale production techniques and facilities are also essential. Hofstein and Chapple (1999) described the commercialization of *Ampelomyces quisqualis* (AQ10) for the control of powdery mildews (PMD) on grape vines. They found that AQ10 did not give acceptable level of control, once the PMD reaches a certain level of infestation (> 30%). The product, therefore, has been presented as a preventive treatment for PMD, if applied when visible incidence is near zero. Based on the experience of developing *A. quisqualis* into a commercial product, they concluded that a candidate biofungicide should be thrown into an authentic commercial situation at an early stage of development to learn about its attributes. Also, that a biofungicide can not combat devastating fungal pathogens causing disease such as PMD as a stand-alone treatment and must be offered as a component within IPM systems. Brannen and Kenney (1997) have also attributed the success of *Bacillus subtilis* strain GBO 3 (Kodiak) in the cotton market to its integration with chemical fungicides. Though biological control agents may have the possibility of replacing chemical fungicides in future, in the near-term use of biologicals require integration with chemical fungicides.

4   Commercialization of Bioherbicides

The whole function of biological weed control is to push the disease process by tipping ecological balance in favor of the pathogens. This can be achieved by applying the pathogen in a viable form, at inoculum levels high enough to initiate an infection and by manipulating the micro-environment for a long enough period of time to make sure that an infection gets to the point that may perpetuate differences. However, an approach directed at finding a unique pathogen that have all the inherent properties necessary to support a commercial product may not be successful. On the other hand, developing technology that can modify the micro-environment and/or enhances the survival of organisms selected to function as a biocontrol agent in a market suited for biological organism may give more dividends.

Fungal pathogens require some periods of dew on the surface of the plant for the spores to germinate and infect the plant. The dew period requirements may vary for at least 8 hours to as much as 72 hours. As natural environment do not provide more than a few hours of dew, a great deal of inconsistency is found in the field trial of the fungi. Thus development of technology to the point that it liaise proper manipulation of the micro-environment, would be a critical requirement for successful commercialization. Most applications of bioherbicides in the field provided little or no help to the organism in serving agricultural function, they are being expected to provide.

Logical bioherbicide tracks are normally considered to be weeds which escape chemical controls, weeds which escape cultural controls, weeds in environmentally sensitive settings and weeds in organic cropping systems. Ideally a proper track weed would be an agricultural system that would permit some environmental modifications associated with an economic incentive. Turf, for example, is generally located in environmentally sensitive locations and irrigation is part of management system. Thus moving a pathogen into the weed would not be a problem. On the other hand, Charudattan (1988) noted that Collego® was an attractive candidate for development because the chemical herbicides used against northern jointvetch were banned in the market place for the herbicides.

Cross and Polonenko (1996) cited an important lesson from the commercialization experience of BioMal®. According to them for commercial success of biocontrol organisms, a change of focus from product-driven research and development to market-driven research and product development is necessary. This requires accurate forecast of customer demand for the biocontrol product, total market potential and determination of the needs and benefits that the product will satisfy.

Bioherbicides also should be formulated for pathogen stability, persistence, efficacy and application. In most cases, shelf life of at least 18 months is critical to successful commercialization. However, chlamydospores of *Phytophthora palmivora* formulated as DeVine® by Abbott rapidly lost viability in storage. Therefore, the product was produced and distributed only after orders were placed (Kenney, 1986). Refrigeration of other formulated fungal herbicides appears critical to storage stability.

5   Commercialization of Bioinsecticides

In order to be successful commercially, the bioinsecticides have to directly and effectively compete with chemical insecticides in the world market. Bioinsecticides surpass the performance of chemical insecticides only when the target pest is resistant to the chemical but susceptible to bioinsecticide. For example, Colorado potato beetle is resistant to a variety of chemicals but it can be effectively controlled by *Bacillus thuringiensis*. The activity of some traditional bioinsecticides has been enhanced using modified formulations but this has not resulted in increasing the spectrum of pest control or level of activity to match the activity of chemical insecticides. Essentially, significant advances in activity, persistence and delivery of bioinsecticides will be necessary to equal their performance to chemical insecticides.

In analysing scientific and business considerations to develop a pesticide discovery, it has been found necessary that the product be patentable, in addition to being competitive and having a wide range of pest control (Stowell, 1993). To cite example, *Autographa californica* nuclear polyhedrosis virus (AcNPV) has been found to provide effective biological control for many caterpillar pests similar to *Bt* bioinsecticides. It has also been proven safe in extensive toxicology tests and has been demonstrated to be effective in field performance evaluations. Unfortunately, AcNPV was not patented after its discovery. The lack of intellectual property protection and no competitive advantage as compared to other biopesticides prevented development of this technology.

Many major companies in the United States are involved in developing bioinsecticides. Most agrochemical companies are mainly working on *Bt*. This technology is well established and it does not require a high investment. Other companies are involved in nematode, virus and fungus technologies. The progress in mass production through fermentation (nematodes, *Bt*) and generations *in vivo* (baculoviruses) were instrumental in marketing these products. Research efforts are in progress to develop and optimize the *in vitro* production process for baculoviruses and fungi.

5.1   Bacillus thuringiensis

Bacillus thuringiensis is the most widely used biological insecticide. The *Bt* technology has been developed to the point that it's capable of protecting crops from insect damage. There are persuasive arguments that *Bt* endotoxins should be regarded as chemicals. *Bt* has evolved to produce large quantities of crystal proteins, making it a logical host for developing improved Cry bioinsecticides. It is now accepted that engineered forms of the Cry proteins showing improved potency or yield, regardless of their host, make Cry bioinsecticides a more attractive and practical alternative to synthetic chemical control agents. However, the introduction of *Bt* transgenic plants will have an impact on the market share of *Bt* sprays, especially in the cotton market.

Farmers with high hopes of controlling destructive insect pests have greeted the new *Bt* products, both plants and sprays, with enthusiasm. The market is likely to grow rapidly as an increasing number of genetically engineered pesticides reach the market and pressure mounts to replace more traditional toxic chemicals.

A comparative economic analysis of a *Bt*-based product and standard chemical products has clearly brought out that in specific situations, even non-engineered *Bt* products perform comparable to or better than chemical standard approach. In a 2-year long study on fresh-market tomato plantings employing separately a *Bt*-product Javelin and methomyl and permethrin, net profits were found to be equal to or better for *Bt* product application than chemical standard approach. Other long-term benefits included reduced frequency of chemical application and, negligible effect on natural enemies (Trumble et al. 1994).

5.2   Baculoviruses

It has been believed that recombinant baculovirus research would result in a unique, competitive insecticide that would control economically important insects with no effect on beneficial species. It could be safely applied to vegetables up to harvest, demonstrate no adverse ecological effects and present no toxicological hazards to mammals. DuPont and American Cyanamid, both have recently tested recombinant nucleopolyhedro viruses (NPVs) in selected cotton and vegetable growing areas of the United States against cabbage looper, cotton bollworm, tobacco budworm, beet armyworm, and diamondback moth. In anticipation of the wide-scale use of recombinant NPVs for the control of lepidopterous pests, effective control strategies are being formulated and tested to ensure the success and acceptance of this new technology. DuPont is taking its new scorpion-loaded virus on the road. This material combines a scorpion venom gene with a baculovirus *Autographa californica* (McCutchen and Flexner, 1999).

It is stipulated that a genetically engineered virus would need to achieve sales of $40 million in order to justify an R&D expenditure of $15 million required. Zeneca, a British life sciences company, terminated a project on recombinant baculoviruses as they believed that many issues remained that must be grappled with and solved before sufficient balance is achieved between the efficacy the growers are happy with and a practical ability to make biopesticides available to these growers (Broadhurst, 1998). The business objectives of Zeneca for genetically engineered baculoviruses were outlined as follows:

(a)   The genetically engineered baculoviruses to have an adequate effect on lepidopteran pests in cotton, vegetables and other crops such as corn.

(b)   The achievable effect is to be equivalent to chemical standards on the least effective end of the spectrum for control of target pests, e.g. thiocarb in cotton.

(c)   A method to produce sufficient quantities of the engineered virus with consistent quality and cost effectively. This requirement was believed to stipulate *in vitro* production methodology with many associated challenges. For example, use of insect cell cultures on large scale under sterile conditions, chiefly driven by an assumption that traditional *in vivo* methods would not be applicable when the larvae died much more quickly.

(d)   Robust formulation technology associated with the formidable task of working with biological materials.

The recombinant baculovirus products have yet to be commercialized but an insight into their deployment scenarios is available. These could be utilized for control of lepidopteran pest species as biocontrol agents in rotational-use strategies, to relax selection pressure exerted by synthetic insecticides or even by *Bt* in pest populations, for use in combination with certain chemical insecticides; and as augmentation of transgenic crop varieties and establishment of binary transgenic plant resistance/biocontrol systems (Treacy, 1999).

5.3   Fungal Bioinsecticides

Fungi are potential bioinsecticides. Their use is a commercial reality and there are a number of fungi such as *Beauveria* and *Metarhyzium* that are widely used, especially in green houses. Mycotech Corporation has registered several formulation of *B. bassiana* strain GHA (BotaniGard®) targeting the silverleaf whitefly, thrips and aphids. As silverleaf whitefly inhabit primarily lower surface of plant leaves, maximum coverage of leaf undersides is obtained by spraying upward from below canopy level. For commercialization, the critical need for intensively replicated field experiments and large scale trials under commercial production conditions can not be overstated. Technological breakthroughs will also be needed, especially in production and formulation, to extend their uses to large acreage areas (Georgis, 1996). Additional research is needed especially with the aim of improving product efficacy and consistency under variable field conditions.

6   Commercialization of Pest-Resistant Crops

The expression of *Bt* genes in plants has provided a widespread application of genetic engineering by shifting the paradigm of external spraying of pest control agents to internalizing this activity within the plant. This approach has provided advantage of effective and environmentally acceptable pest control in large-acreage crops such as cotton and corn and eliminated or reduced the need of repeated foliar application of pest control agents.

6.1   Insect-Resistant Cotton – A Case Study

An account of Monsanto's journey through development of Bollgard® cotton from laboratory to market place, spanning a nearly decade of efforts, has been provided by Perlak and Fischhoff (1996), Sims (1996) and Goldberg and Tasker (1997) (Figure 11.2).

6.1.1   Insect-Resistant Plants - Strategic Positioning

Based on successful research in introducing and expressing the protein-toxin genes in tobacco and tomato plants and finding them insecticidal in laboratory and green house bioassays, the identification of a key crop for the first introduction of insect-resistant plants was considered a necessity by Monsanto. The factors in choosing cotton as a focus crop included technical feasibility of transformation, high value crop to justify the effort and consumer acceptance of the technology. It is a dicotyledonous plant and has been successfully regenerated and transformed by *Agrobacterium*-mediated transformation. Cotton was one of the most chemical-intensive crops grown in approximately 12.5 million acres in the United States alone. Despite heavy chemical use, on average, farmers still lost up to 20% of their crop to insect pests. Insect damage remained a serious problem with the bollworm/budworm complex consisting of tobacco budworm (*Heliothis virescens*), cotton budworm (*Helicoverpa zea*) and pink bollworm (*Pectinophora gossypiella*). These are all sensitive to *Bt* endotoxins.

6.1.2   Technical Advances Required

While technical feasibility of cotton regeneration and transformation was known, its efficiency was low. A commitment to efficient plant transformation was required followed by screening of close to 1000 transgenic cotton plants in

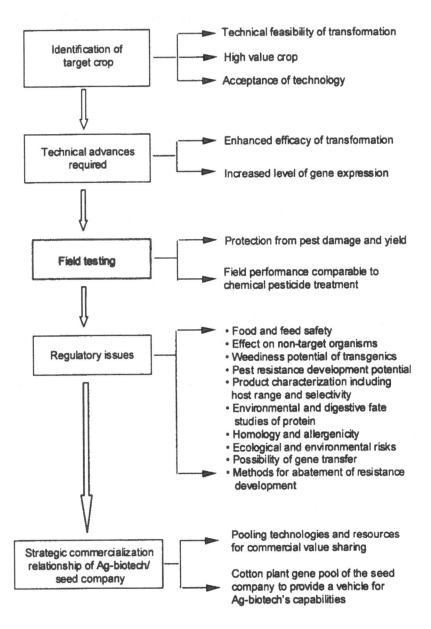

Figure 11.2   Stages of laboratory to commercialization of *Bt*-cotton

a green house test with cotton bollworm for identification of best plants for back-crossing programs into commercial "elite" lines. New approaches were initiated for increased expression of *cryIAb* protein genes in genetically modified cotton. Site-directed mutagenesis was employed to partially modify the gene by removing the DNA sequences predicted to inhibit efficient expression. This gene increased the expression in plants by 100-fold compared to the wild-type (WT) gene. Another approach used a fully modified synthetic gene that encoded a protein identical in amino acid sequence to the WT gene. Consequent to modification of DNA sequence, expression levels in plants increased up to 1000-fold over levels seen with wild-type genes in plants. This directly correlated to increased insecticidal activity. The increased expression was generic across several plant species including tobacco, tomato, cotton and corn and was obtained with two distinct genes, *cryIAb* and a modified truncated *cryIAc* gene. The modified genes were introduced in cotton resulting in very high levels of insect-control protein gene expression.

6.1.3   Field-testing of Insect-Resistant Cotton

The improved transformation system and the increased levels of target gene expression were combined to generate plants suitable for field-testing. Beginning from the day a cotton transformation was initiated to the day the first preliminary results are obtained from the field, it took over two and half years. The plants demonstrated excellent protection from lepidopterous pests. The protection obtained was found comparable to that obtained using weekly treatment of chemical insecticides for lepidopteran control on non-transformed plants. The yield of cotton from transgenic cotton in the field tests was also comparable to the insecticide-treated non-trangenic cotton.

6.1.4   Regulatory Issues

The *Bt* proteins are generally regarded safe; the issues related to food and food safety, effects on non-target organisms, weediness potential of transgenics and also the potential for pest resistance development were investigated (Sims, 1996). Besides, when one deals with microprojectile bombardment of a cell for constructing a recombinant DNA system, it is difficult to get a single copy of the gene go into the plant cell cleanly. Often, either multiple copies or partial copies result (see figure 3.1) and these were required to be defined and described to the regulatory agency, EPA. The product characterization included host range studies and verification that the proteins are indeed selective to target pest species. A homology search was carried out and allergenicity was determined to find if the *Bt* protein is like any known toxins or allergens. Digestive fate studies were performed *in vitro* using gastric- and intestinal-fluids at different pHs to be sure that protein digested.

Ecological and environmental risks were evaluated for adverse effects on non-target organisms: avian, mammalian, aquatic, beneficial insects and endangered species. These tests were unique to the crop and pests involved. For example, for Bt-corn, tests were conducted on the potential effects on fish because field corn may be manufactured into commercial fish food. Also investigated were the issues related to genes such as possibility of their transfer to other plants or problems caused in the plant itself. Environmental fate of the protein was assessed to see if it accumulated and became a problem to non-targets, especially in the soil, and to honeybee, ladybeetle, hymenoptera (parasitoids) and lacewing. Methods for abatement of resistance development were also investigated.

6.1.5   Pooling Technologies and Resources for Commercial Value Sharing

Monsanto needed a seed company to provide the vehicle for taking their genetic engineering capability to market. Delta and Pine Land (D&PL), a Mississippi-based cotton seed market leader, stood to have an industry advantage if it could be the first and possibly the only company to market a transgenic cotton product that worked. D&PL managed one of the largest cotton plant gene pool and needed to capture some value for the use of its germplasm as a transport agent for Monsanto's biotechnology. It had a reputation for consistently providing the highest-quality cotton seed varieties that gave the farmer the right yield, optimum maturation cycle and fiber quality characteristics.

However, there was an uncertainty of the value of *Bt* cotton seeds use that would be created for those involved in the value chain. These included the farmers, the seed companies, the technology provider, the distributors and the agrochemicals companies who stood to loose money through decreases in sales. It was clear that there would be a complete restructuring of the industry before the allocation of that value was fully understood.

Monsanto and D&PL drew the first contract in 1992, that was subsequently amended in 1994 and 1996. The February 1996 revision made provision for Monsanto to license its technology directly to the farmer and the retailer being required to attend a training session and to sign a contract. The goal was to communicate the value of newly purchased seed with biotech traits and dissuade the farmer from saving his *Bt* cottonseed with a high penalty if found guilty of saving it.

The direct licensing agreements as a marketing process seemed to be phenomenal success and added considerable value for both Monsanto and D&PL. This generated an extensive database of the cotton farmers' businesses through the grower licensing agreements and established a precedent of direct marketing to the consumer in the agriculture industry. It provided an opening for Monsanto, in the future, and with other biotech products, an ability to implement a strategy of mass customization, marketing directly to farmer, constructing individual deals (by product and geography) that optimized value. On the other hand, from the seed company perspective, they made more money per acre from the transgenic cotton fee than on the non-transgenics seed margin. They could also demonstrate the 'leap of faith required by farmers in the product they sold based on their reputation of quality seed company'.

Lessons from the commercialization of transgenic cotton would predict that anticipated complexities of restructuring must include providing patent protection, a system for collection of royalties and changes in current farm and industry structures, including partnerships that incorporate access to enabling technologies into traditional seedstock companies (Goldberg and Tasker, 1997).

Similar to Monsanto joining hands with D&PL for the development of *Bt* cotton lines, Mycogen who developed new corn hybrids resistant to European corn borer, found that it had neither the time nor the resources to go it alone in developing new engineered plant varieties and sought partnership to pool its technology and resources. Mycogen collaborated with Pioneer Hybrid for development of insect resistant crops based on *Bt*. Such collaboration by combining complementary areas of expertise are expected to allow companies to make rapid progress in development of new agricultural products.

7   Concluding Remarks

There are a number of critical factors that contribute equally to the ultimate success of a product development and commercialization program of biologicals (Leggett and Gleddie, 1995). All of these are broadly in conformity with the experience gained from already commercialized biopesticides and can be summarized as follows:
1)   Sufficient customer demand and market size to ensure a financial return on the investment in research and development in a reasonable time period;
2)   a cost-effective manufacturing process;
3)   stable, effective and easy-to-use formulations;
4)   product compatibility with existing distribution systems and agricultural practices;
5)   highly efficacious, reliable and reproducible field performance by the formulated product;
6)   the ability to obtain patent protection and product registrations and
7)   sufficient sales revenues generated to support continued marketing and promotion, and to continue further product developments.

Contact chemical agents have conditioned growers to evaluate control based on quick knock down. There are few sights more gratifying to a grower than that of yield-threatening pests perishing shortly after a chemical pesticide is applied. Biologicals work more slowly, requiring different, more complicated criteria such as yield for measuring effectiveness. Therefore, a change of mindset is also essential for encouraging broader use of biopesticides. This transition will require growers to be better educated about biological control technologies through cooperative extension services. For this to happen, extension personnel who have immense knowledge base to support the use of chemical pesticides will have to be provided with new knowledge for biological agents for transmittal to growers.

Commercialization of genetically engineered crops that began in 1996 with a ready acceptance by farmers attained an impressive and rapidly expanding growth. Farmers accepted *Bt* corn, for example, much faster as compared to hybrid corn. However after four years of explosive growth, by late 1999, increasing consumer concerns towards environmental impact and food safety had resulted in a backlash. Environmental issues related to biodiversity, gene pollution, gene transfer, and insect resistance and food safety issue related to antibiotic resistance, allergic reactions and health effects, had created controversies and distrust. Many of these issues remain unresolved, as no unequivocal answers were available yet. If perceptions create reality in the market place, then reports such as transgenic potatoes fed to rats causing thickening of the walls of part of their digestive tract (Ewen and Pusztai, 1999) and laboratory mortality of monarch butterfly larvae fed on pollen from *Bt* corn (Losey et al., 1999), have not helped the matters. This may have an effect of stunting the growth of genetically modified crops market, at least temporarily. By the end of 1999, Monsanto had agreed to modify their gene technology licensing agreements with Delta & Pine Land for insect-resistant and herbicide-tolerant crops and abandoned the plans of it's acquisition, setting it free to talk to other potential suitors. Subsequently, Monsanto has merged with Pharmacia & Upjohn with a speculation of its agribusiness would be spun off in a public offering.

The market appears to be driven by perceptions that determine consumer choice rather than rationale in terms of the safety alone. The agronomic or "input traits" designed with an eye to farming practices and pesticide use are visible only to farmers. Positive environmental effects, such as decreased pesticide use are seen beneficial but the consumer perceives real value in terms of lower prices and/or better food quality. Some of the surveys, however, give hope that acceptance will grow as consumer knowledge about biotechnology increases (Thayer, 1999). Particularly, advancements towards next-generation genetically modified crops with improved "output traits", such as tangible benefits for health and nutrition might even create demand and be crucial in winning over the consumers. In the end, consumers are the ones who will make the choices and will ultimately drive the market, not technology. But, more and better science-based information must be made available to help them with these choices.

References

1.   Brannen, P. M. and D. S. Kenney, 1997. "Kodiak® – a successful biological-control product for suppression of soil-borne plant pathogens of cotton", *J. Ind. Microbiol. Biotechnol.*, 19, 169 – 171.

2.   Broadhurst, M. D., 1998. "Biopesticides: From disillusionment to integrated crop management- A large company perspective", Abstracts, *Alternative Paradigms for Commercializing Biological Control Workshop*, Rutgers University, New Brunswick, NJ, Internet: http://www.rci.rutgers edu/~insects/broadsum.htm.

3.   Cross, J. V. and D. R. Polonenko, 1996. "An industry perspective on registration and commercialization of biocontrol agents in Canada", *Can. J. Plant Pathol.*, 18, 446-454.

4.   Charudattan, R., 1988. in *Fungi in Biological Control Systems*, M. N. Burge, ed., Manchester University Press, Manchester, pp. 86-110.

5.   Ewen, S. W. and A. Pusztai, 1999. "Effect of diets containing genetically modified potatoes expressing *Galanthus nivalis* lectin on rat small intestine", *Lancet*, 354, 1353-1354.

6.   Froyd, J. D., 1997. "Can synthetic pesticides be replaced with biologically-based alternatives? – an industry perspective", *J. Ind. Microbiol. Biotechnol.*, 19, 192-195.

7.   Gardner, W. A. and C. W. McCoy, 1992. "Insecticides and herbicides", in *Biotechnology of Filamentous Fungi: Technology and Products*, D. B. Finkelstein and C. Ball, eds., Butterworth – Heinemann, Stoneham, MA, USA, pp. 335-359.

8.   Gaugler, R., 1997. "Alternative paradigms for commercializing biopesticides", *Phytoparasitica*, 25, 3.

9.   Georgis, R., 1996. "Present and future prospects of biological insecticides", *Conference on Biological Control*, April 11-13, 1996, Cornell Univ., Internet: http://www.nysaes.cornell.edu/ent/bcconf/talks/ georgis.html.

10.   Goldberg, R. A. and C. Tasker, 1997. "Delta & Pine Land: Measuring the value of transgenic cotton", *Harvard Business School Case Study* 9-597-005 (Revised), Harvard Business School Publishing, Boston.

11.   Harman, G. E., 1996. "*Trichoderma* for biocontrol of plant pathogens: From basic research to commercialized products", *Conference on Biological Control*, Cornell Univ., April 11-13, 1996. Internet: http://www.nysaes.cornell.edu/ent/bcconf/talks/harman.html.

12.   Hofstein, R. and A. Chapple, 1999. "Commercial development of fungicides", in *Biopesticides: Use and Delivery*, F.R. Hall and J.J. Menn, eds., Humana Press, Totowa, NJ, pp.77-102.

13.   Kenney, D. S., 1986. "DeVine® - The way it was developed - an industrialists' view", *Weed Sci.*, 34, 15-16.

14. Legget, M. E. and S. C. Gleddie, 1995. "Developing biofertilizer and biocontrol agents that meet farmers' expectations", *Adv. Plant Pathol.*, 11, 59-74.

15. Losey, J. E., L. S. Rayor and M. E. Carter, 1999. "Transgenic pollen harms monarch larvae", Nature, 399, 214.

16. McCutchen, W. F. and L. Flexner, 1999. "Joint actions of baculoviruses and other control agents", in *Biopesicides: Use and Delivery*, F.R. Hall and J.J. Menn, editors, Humana Press, Totowa, N.J., pp. 341-355.

17. Perlak, F. J. and D. A. Fischhoff, 1993. " Insect-resistant cotton: From laboratory to the marketplace", *Advanced Engineered Pesticides*, L. Kim, ed., Marcel Dekker, New York, NY, pp 199-211.

18. Shimoda, S.M. 1997. "Challenges of commercializing biopesticides in a more competitive marketplace", *Second Annual Conference on Biopesticides and Transgenic Plants: New Technologies to Improve Efficacy, Safety and Profitability*, Washinton D.C., Jan. 27-28, 1997.

19. Sims, S., 1996. "Development and commercialization of insect resistant transgenic crops", *Conference on Biological Control*, Cornell Univ., April 11-13, 1996, http://www.nysaes.cornell.edu/ent/bacconf/talks/sims.html.

20. Stewart, R. S., 1999. "Toward one market: harmonising efforts between the US and Canada may be threatened by FQPA and GMOs", *Farm Chemicals international*, 13(2), 25-26.

21. Stowell, L. J., 1993. "Factors influencing acceptance and development of biopesticides", in *Advanced Engineered Pesticides*, L. Kim, ed., Marcel Dekker, Inc., New York, pp. 249-260.

22. Sutton, J. C., 1995. "Evaluation of microorganisms for biocontrol: *Botrytis cinerea* and strawberry, a case study", *Adv. Plant Pathol.*, 2, 173-190.

23. Thayer, A. M., 1999. "Ag biotech food: risky or risk free", Chem. & Engg. News, 77(44), 11-20.

24. Treacy, M. F., 1999. "Recombinant baculoviruses", in *Biopesticides: Use and Delivery*, F. R. Hall and J. J. Menn, eds., Humana Press, Totowa, NJ, pp. 321-340.

25. Trumble, J. T., W. G. Carson and K. K. White, 1994. "Economic analysis of a *Bacillus thuringiensis*-based integrated pest management program in fresh-market tomatoes", *J. Econ. Entomol.*, 87(6), 1463-1469.

26. Wilson, C. L. and M. A. Jackson, 1997. "Commercializing of biologically-based technology for the control of plant pests", *J. Ind. Microbiol. Biotechnol.*, 19, 156.

27. Wood, H. A., 1995. "Development and testing of genetically improved baculovirus insecticides", in *Baculovirus Expression Systems and Biopesticides*, M. L. Shuler, ed., Wiley-Liss, New York, pp. 91-102.

28. Zorner, P.S., S. L. Evans and S. D. Savage, 1993. Perspectives on providing a realistic technical foundation for the commercialization of bioherbicides, in *Pest Control with Enhanced Environmental Safety*, S.O. Duke, J.J. Menn and J.R. Plimmer, eds., American Chemical Society, Washington, DC, pp. 79-86.

Glossary, and
Product, Manufacturer,
Pathogen and Subject Indices

Glossary

Actinomycete
(Literally
"ray fungi")

Filamentous bacteria that have sometimes been classified as a fungi imperfecti. The name is often used to refer specifically to those species that form mycelium.

Adjuvant

Material added to improve some chemical or physical property (e.g., of a plant protectant) such as solvents, diluents, carriers, emulsifiers, stickers or spreaders.

Agroecosystem

A relatively artificial ecosystem in an agricultural field, pasture, or orchard.

Allele

Any of one or more alternative forms of a given gene; both (or all) alleles of a given gene are concerned with the same trait or characteristic.

Antagonism

An ecological association between organisms in which one or more of the participants is harmed or has its activities limited.

Antagonist

An agent or substance that counteracts the action of another.

Antibiosis

An association between two organisms in which one harms the other.

Apothecium

An ascus-bearing structure (ascocarp) in which the ascus-producing layer (hymenium) is not covered by fungal tissue at maturity.

Appressorium

An enlargement on a hypha or germ tube that attaches itself to the host before penetration takes place. (Pl. appressoria.)

Arthropod

A phylum or division of the animal kingdom: include insects; spiders and crustace.

Autosome

All chromosomes except the sex chromosomes.

Bacteriophage

Virus that lives in and kills bacteria. Also called phage.

**Biological
control**

The deliberate use by humans of one species of organism to eliminate or control another.

Biotroph

An organism that derives nutrients from the living tissues of another organism (its host).

Blastospore

A spore that arises by budding, as in yeasts.

Brush border membrane	A superficial protoplastic modification in the form of filiform processes or microvilli present on certain absorptic cells in the ntestinal epithelium and the proximal convolutions of nephrons.
Chlamydospore	A loosely used term usually applied to a thick walled, asexually-produced resting spore formed by certain fungi; the term is sometime used to refer specifically to the teliospore of the smut fungi.
Cloning	An *in vitro* procedure in which a particular sequence of DNA (e.g. a gene) is reproduced in large amount by inserting ('splicing') it into a suitable replicon (the vector or cloning vector), introducing the resultant recombinant (hybrid) molecule into a cell in which it can replicate and finally growing the cells in culture.
Columnar epithelium	Epithelium distinguished by elongated, columnar or prismatic cells.
Cytolytic	Able to lyse cells.
Damping-off	Collapse and death of seedling plants resulting from the development of a stem lesion at soil level.
Dicotyledon (Dicot)	Plant whose seeds have two cotyledons or seed leaves, such as beans.
Deuteromycetes	A large miscellaneous artificial group of fungi in which sexual reproduction does not occur or has not been found, contains most of the wilts and some damping off fungi.
Endogenous	Growing throughout the substance of the stem, instead of by superficial layers.
Endoparasite	Any of the various parasites which lives within the body of its hosts.
Endophyte	An organism parasitic partly or wholly within a plant.
Endospore	A type of spore formed intracellularly by the parent cell or hypa and are formed under conditions of nutrient limitation.
Endotoxin	A toxin produced within an organism and liberated only when the organisms disintegrates or destroyed.
Entomogenous	Growing in or on insects.
Entomopatho-genic	Insect-attacking organism.

Entomophagous insect	Any insect that eats other insects.
Entomophagous parasite	An insect or fungus that parasites insects.
Enzyme	A large complex protein molecule produced by the body that stimulates or speeds up various chemical reactions without being used up itself; an organic catalyst.
Epiphyte	A plant growing on another (usually not fed by it) or a micro-organism living on the surface of plants in a non-parasitic relationship.
Epithelium	A primary animal tissue, distinguished by cells being close together with little intercellular substance covers free surfaces and lines body cavities and ducts.
Epizootic	A disease outbreak within an insect population.
Erythrocyte	A type of blood cell that contains a nucleus in all vertebrates. Also known an red blood cell.
Exogenous	1) Produced on the outside of another body, 2) produced externally, as spores on the tips of hyphae, 3) Growing by outer additionals of annual layers, as the wood in dicotyledons.
Exospore	A type of spore formed form the parent organism by budding or by septum formulation and fission.
Exotoxin	A soluble toxin excreted by specific bacteria and absorbed into the tissues of the host.
Expression	In genetics, manifestation of a characteristic that is specified by a gene. With hereditary diseases, for example, a person can carry the gene for the disease but not actually has the disease. In this case, the gene is present but not expressed. In industrial biotechnology, the term is often used to mean the production of a protein by a gene that has been inserted into a new host organism.
Facultative	Designating an organism which is capable of living under more than one condition: e.g. a saprophyte and as a parasite; as an aerobic or anaerobic organisms.
Genetic modification	A process that results in a change in the genetic make up of a population. Methods of inducing genetic modification include conjugation, *in vivo* rearrangements of transposable elements, *in vitro* gene recombination techniques, protoplast fusion, the use of mutagens, hybridization by normal breeding techniques.

Genome	The total hereditary material of a cell, comprising the entire chromosomal set (hence of genes) found in each nucleus of a given species.
Genotype	The genetic constitution of an organism.
Goblet cell	A unicellular, mucus-secreting intraepithelial gland that is distended on the free surface.
Gram stain	A staining method devised by Danish physician Hans Grams to aid in the identification of bacteria. Bacteria either resist discolorization with alcohol and retain the initial deep violet stain (Gram positive) or can be decolorized by alcohol and are stained with a contrast stain (Gram negative).
Hemocoel	An expanded portion of the blood system in arthropods that replaces a portion of the coelom.
Hemolymph	The circulating fluid of the open circulatory systems of many invertebrates.
Hemolysis	The destruction of red blood cells and resulting escape of hemoglobin.
Heterozygous	An animal that carries genes from two different characters (Impure).
Homologous recombination	Occur only between two sequences that have fairly extensive regions of homology.
Hyperparasite	An organism that is parasitic on another parasite.
Hypha	The simple or branched thread like filaments that compose the web like mycelium of fungi.
Inoculum	That portion of individual pathogen or its parts that can cause infection and that is introduced into or transferred to a host or medium.
Intraoccular	Within the globe of the eye.
Intraperitoneal	Within the cavity of the body that contain the stomach and intestines.
In vitro (literally in glass)	Frequently in the sense "not under natural conditions" e.g. in the laboratory; in experimental culture (of chemical reactions) not in living cells.
In vivo	In the living organism.

Lesion	A localized area of discolored, diseased tissue.
Lethal concentration	The lethal concentration (written as LC_{10} or LC_{50} or LC_{100} or any percentage) median is the parts per million (ppm) or parts (LC) per billion (ppb) of toxicant in water or air which kills 10 or 50 or 100% respectively of the target species in a 24-hour period.
Lethal dose (LD)	The lethal dose (written LD_{10}, LD_{50}, LD_{100} or any percentage) median is the milligrams of toxicant per kilogram of body weight that kills 10, 50 or 100% of the target species.
Lysis	The enzymatic dissolution of all or part of a uni- or multi-cellular structure or dissolution of a phage-infected bacterium.
Metamorphosis	A process by which an organism changes in form and structure in the course of its development, as many insects do.
Microvillus	One of the filiform (threadlike or filamentous) processes that form a brush border on the surfaces of certain specialized cells, such as intestinal epithelium.
Mildew	A fungal disease of plants in which the mycelium and spores of the fungus are seen as a whitish growth on the host surface.
Mold	Any profuse or woolly fungus growth on damp or decaying matter or on surfaces of plant tissue. Blue mold or green mold is caused by *Penicillium* spp. and grey mold, by *Botrytis cinerea*.
Monocotyledon (Monocot)	Plant having a single cotyledons or seed leaf such as corn.
Mutant	A variant, different genetically and often visibly from its parent or parents and arising rather suddenly or abruptly. Of an organism, population, gene, chromosome, etc.: Differing from the corresponding wild type by changes in one or more loci.
Mutation	A sudden random change in the genetic material of a cell. A stable, heritable change in the nucleotide sequence of a genetic nucleic acid (DNA, or RNA in viruses, viroids, etc) typically resulting in the generation of a new allele and a new phenotype. Mutation can occur naturally or can be induced by radiation (X-ray, gamma ray or thermal neutrons) or chemically.
Mycelium	The hypha or mass of hyphae that makeup the body of a fungus. A mass of interwoven filamentous 'threads' that make up the vegetative part of a fungus. Mycelia (pl.).
Mycoparasite	A fungus parasitic on other fungi.

Mycosis   An infection by a parasitic fungus, or a disease so caused.

Mycotoxins   Chemical substances produced by fungi that may result in illness
                        and death of animals and humans when food or feed containing them
                        is eaten.

Necrotroph   An organism that derives nutrients from dead plant or animal tissues,
                        whether or not it is responsible for the death of those tissues.

Neurotoxicity   The state or condition of being poisonous to the brain and nerves of
                        the body.

Obligate   Description of an organism that requires a particular set of
                        environmental conditions or nutrient for growth or survival. For
                        example, an obligate aerobe will only grow in the presence of
                        oxygen and an obligate anaerobe's growth will be inhibited or will
                        be killed in the presence of oxygen.

**Obligate   A parasite that in nature can grow and multiply only on or in living
parasite**   organisms.

Oospore   A thick-walled spore produced by sexual reproduction in downy
                        mildews and related fungi.

Parasite   An organism that lives at least for a time on or in at the expense of
                        living animals or plants.

Pathogen   Any microorganism which by direct interaction with (injection of
                        another organism (by convention, a multi-cellular organism) cause
                        disease in that organism.

Pathogenicity   The ability of a pathogen to cause disease.

**Peritrophic   A chitinous tube free within the cavity of the midgut of insects
membrane**   which separates food from the lining epithelium.

Phenotype   Individuals of the same phenotype look alike but may not breed
                        alike.

Phylloplane   The surface(s) of a leaf.

**Phytopatho-   Term applicable to a microorganism that can incite disease in plants.
genic**

Phytotoxic   Toxic to plants or plant growth.

Plasmid   A circular piece of DNA found outside the chromosome in bacteria.
                        Plasmids are the principal tools for inserting new genetic
                        information into the microorganisms or plants.

Plasmid vector	A plasmid used as cloning vehicles or vectors for the introduction of foreign DNA-containing genes that do not normally occur in the host cell.
Polyphagous parasites	A parasite which is capable of parasitizing considerable number of host species.
Propagule	The part of an organism that may be disseminated and reproduce the organism.
Proteinase	An enzyme that digests proteins and acts directly on the native proteins in their conversion to simpler substances.
Proteolytic Enzyme	Any enzyme that catalyzes the breakdown of protein.
Pycnidium	A closed sporocarp, usually opening by a pore, that contains a cavity bearing conidia.
Recombinant DNA technology	The technique of isolating a gene and inserting it into the DNA of another organism, also called genetic engineering, gene splicing or genetic modification.
Restriction Enzymes	Enzymes that can be used as molecular 'scissors' to cut DNA into reproducible fragments at specific sites. There are many different kinds of restriction enzymes and a large number of them have been carefully catalogued according to the point at which they will cut a DNA molecule. By selecting the appropriate restriction enzyme, a given DNA molecule can be cut at a useful site.
Rhizosphere	The soils region on and around plant roots. The root zone, the area where microorganisms are most active in increasing the availability of nutrients for plants.
Rot	The softening, discoloration, and often disintegration of a succulent plant tissue as a result of fungal or bacterial infection.
Rust	Any of various plant diseases causing to form rust-colored spores on affected plants.
Saprophyte	Plants, including certain bacteria and molds, capable of obtaining nutrients and energy from dead organic matter.
Sclerotium	Hard, resistant, multicellular resting body, that under favorable conditions can germinate to produce mycelium or sexual or asexual fruiting bodies. Sclerotia (Pl.).
Senescence	The process of growing old. Decline or degeneration, as with maturation, age, or disease stress.

Septicemia   Blood poisoning, a disease condition which results from the presence of toxins or poisons of microorganisms in the blood.

Serotyping   Strains are distinguished on the basis of difference in their surface antigen.

Siderophore   A metabolic product of a fungus (or other organism) which binds iron and facilitates its transport from the environment into its microbial cell.

Site-directed   The process of introducing specific base-pair mutations into a gene.
mutagenesis

Smut   Any of a number of plant diseases characterized by masses of dark, powdery, and sometimes odorous spores (e.g., stinking smut of wheat, common smut of maize).

Species   A unit used in the classification of living organisms. Members of the same species resemble each other closely. In the naming of plant and animals Latin is used. Each kind of plant or animal can be identified by genus (Plural: genera) and species (both singular and plural) eg. generic name (genus) of corn is *Zea* and the species name is *mays*.

Sporangium   A fungus fruiting body that produces non-sexual spores within its external walls.

Spore   A discrete sexual or asexual reproductive unit, usually enclosed by a rigid wall, capable of being disseminated. The reproductive unit of fungi consisting of one or more cells; it is analogous to the seed of green plants.

Strain   A sub-species group of organisms distinguishable from the rest of the species by a heritable characteristic that the individuals in the group have in common.

Stroma   A compact mass of vegetative tissue, sometimes intermixed with host tissue, often bearing sporocarps either within or upon its surface. (Pl. stromata).

Synergism   The association of two or more organisms acting at one time and effecting a change which one only is not able to make. The concurrent parasitism of host by two pathogens or effect of a chemical in which the symptoms or other effects produced are of greater magnitude than the sum of the effects of each pathogen or a chemical acting alone.

Subcutaneous   Situated or occurring beneath the skin.

Toxinomy heirarchy	Division (in botanical schemes) or phylum (in zoolyical scheme), class; order; family, genus, and species.
Transgenic	An organism that has been transformed with a foreign DNA sequence.
Transgenic *Bt* plants	A gene from the *Bt* bacterium is inserted into the DNA of a plant cell. When the cell is re-generated or "grown" into a whole plant, the resultant plant, is resistant to specific insect pests.
Ubiquitous	Occurring everywhere as house flies, weeds.
Virion	The infectious unit of a virus.
Virulent	Capable of causing a severe disease; strongly pathogenic.
Wilt	A plant disease characterized by loss of turgidity, which results in drooping of leaves, stems, and flowers.
Zoospore	An asexual, motile reproductive cell, which, by repeated divisions of the protoplast, develops into a new plant: a means of reproduction of many green algae and lower fungi.
Zooplankton	Tiny animals which drift with the currents.

Product Index
(Product and its manufacturer)

T

V

X

Y

Manufacturer Index
(Manufacturer and its products)

A

Abbott Labs. (Now Valent BioSciences), Chicago IL, USA, 8, 10, 44, 48t, 52, 57, 80, 106, 107t, 108, 111, 112, 113, 114, 115, 125, 152, 213t, 214, 217, 220, 234, 235, 236t, 237, 253 — Dibeta, DiPel, DeVine, Novodor, VectoBac FC, VectoLex, XenTari

AgBiochem, Inc., Orinda CA, USA, 170t, 172 — Galltrol

AgraQuest Corp., Davies CA, USA, 170t, 172, 219, 220t — Lagniex, Serenade

Agricola El Sol, Brazil, 151t — VPN

Andermatt Biocontrol, Switzerland, 151t — Madex 3

AgrEvo, Wilmington, DE, USA, 94t — StarLink

B

Bayer AG, Germany, 219, 220t — BIO 1020

Becker Microbial Products, Inc., Plantation FL, USA, 48t, 57, 103, 104t — $Aquabac_{xt}$, BMP123

Behring AG, Werke, Germany, 151t — Granusal

BioCare Technology, Somersby, Australia, 170t, 172 — Nogall

Biochem Products, 66t — Spherimos

Bio-Innovation AL, Toreboda, Sweden, 174t, 179, 179f — Binab T

Biological Prepn. Berdesk, Russia, 66t, 69, 103, 104t, — Bactoculicide, Spherix

BioWorks Inc., Geneva NY, USA, 173, 174t, 191 — T-22, Bio-Trek 22G, Root Shield

C

Caffaro S.p.A., CeSano Maderno, Italy, 8 — Exobac

CCT Corpn. Carlsbad CA, USA, 171t, 173, 185t,   Deny, Dr. Biosedge
    193, 213t, 214

D

Dekalb Genetic Corporation, Dekalb, Il, USA,   Bt-Xtra
    94t

Dominion Bio-Sciences, Inc., Walnut Creeck,   Leone
    CA, USA, 171t

E

Ecological Labs., Dover, England, 175t, 175   P.g. Suspension, Rot Stop .

Ecogen Inc., Langhorne PA, USA, 10, 79, 79t,   Dagger G, Foil, Lepinox, Raven
    83, 88t, 89, 111, 112, 113, 134, 134f, 135, AQ^{10}, Aspire, Condor, CryMax,
    171t, 172, 183, 183t, 185t, 186, 192, 252   Cutlass,

EcoScience , Worcester MA, USA, 185, 185t,   BioPath, Bio-Save 11, 110 & 1000,
    192, 219, 220t   ESC 170GH

EcoSoil System   Spot-less

Eden Bioscience, Poulsbo WA, USA, 183t   Grey Gold

Encore Technologies, Minnetonka, MN, 213,   Collego
    213t, 214, 217, 218, 220, 253

G

Gustafson Inc., Dallas TX, USA, 170t, 172, 192,   Epic, Kodiak
    252

H

Helena Chemical Co., Memphis, TN, USA, 230   System 3 Seed Treatment

I

IARI, New Delhi, India, 175t, 176   Kalisena

International Mineral & Chemical Corporation,   Sporeine
    4

J

JH Biotech, Ventura, CA, USA, 174t   Promote

K

Kemira Agro Oy, Helsinki, Finland, 171t, 173,   Mycostop
193

M

Makteshim Chemical Works, Beer Sheva, Israel,   Trichodex
182, 183t, 228, 229t

Mauri Foods, North Ryde, Australia, 171t, 172   Conquer

Monsanto Co., St. Louis MO, USA, 92t, 94t,   Bollgard, NewLeaf, NewLeaf Plus,
137, 138, 257   Yield Gard

Mycogen Crop Protection, San Diego CA, USA,   Casst, Mattch, NatureGard, X-PO,
84, 85, 87, 87t, 88t, 94t, 111, 115, 115f, 213,   M-Peril, MVP, MVP II
213t, 215, 218, 247

Mycotech Corp., Butte MT, USA, 139, 219,   Mycotrol-GH, BotaniGard
220t, 256

N

Natural Plant Protection S.A.,Nogueres, France,   Mamestrin, Spodopterin,
151t, 174t, 178, 192, 238   Caprovirusine, Fusaclean,

New Bio Products, Corvallis, OR, USA, 170t,   Norback 84-C
172

Novartis Crop Protection, Greensboro NC,   Attribute, Elcar,
USA, 94t, 151t, 155

Novo-Nordisk Biochem North America Inc.,   Foray 48B, Florbac FC
Franklinton NC, USA, 111, 114t

P

Pharmacia & Upjohn, Kalamazoo, MI, USA,   Collego
213, 213t, 214, 217, 218, 220, 253

Philom Bios, Saskatoon, SK, Canada, 213t, 214,   BioMal
218, 253

Pest and Pathogen Index
Organism (Common name)

A

Subject Index

Printed in the United States
by Baker & Taylor Publisher Services